Traditional Woodwork

Traditional Woodwork

ADDING AUTHENTIC PERIOD DETAILS TO ANY HOME

MARIO RODRIGUEZ

The Taunton Press

Cover photos by Vin Greco

BOOKS & VIDEOS

for fellow enthusiasts

Text © 1998 by Mario Rodriguez
Photos © Mario Rodriguez, except where noted
Illustrations © 1998 by The Taunton Press

Printed in the United States of America
10 9 8 7 6 5 4 3 2 1

The Taunton Press, 63 South Main Street, PO Box 5506, Newtown, CT 06470-5506
e-mail: tp@taunton.com

Distributed by Publishers Group West

Library of Congress Cataloging-in Publication Data

  Rodriguez, Mario, 1950 –
     Traditional woodwork : adding authentic period details to any home
  / Mario Rodriguez.
       p.   cm.
     Includes index.
     ISBN 1-56158-176-3
     1. Woodwork.   2. Architectural woodwork–United States–Designs
  and plans.   3. Architecture, Colonial–United States.   I. Title.
  TT180.R59 1998
  684'.08–dc21   98-24962
                                                      CIP

About Your Safety

Working wood is inherently dangerous. Using hand or power tools improperly or ignoring standard safety practices can lead to permanent injury or even death. Don't try to perform operations you learn about here (or elsewhere) unless you're certain they are safe for you. If something about an operation doesn't feel right, don't do it. Look for another way. We want you to enjoy the craft, so please keep safety foremost in your mind whenever you're in the shop.

In memory of my daughter, Isabel Rose,
whose short life brought so much happiness to my own

ACKNOWLEDGMENTS

I'd like to thank my friends and colleagues who generously gave their time and talents to this projects. They include Rick Anger, Edward Fitzpatrick III, Henry Engelhardt, and the members of The Warwick Historical Society, Warwick, New York. I am grateful to the members of my family, who have always supported and encouraged me in my work, especially my wife, Judy. I would also like to thank Anthony LeClerc, of General Machinery, Drummondville, Quebec, Canada; Historic Housefitters, of Brewster, New York; Jody Garret, of Woodcraft Supply; Wally Wilson, of Veritas; Jean Miskimon, of Eisner & Associates; Frank Cootz, of Ryobi; and my editors, Rick Peters and Ruth Dobsevage, and others at The Taunton Press.

CONTENTS

INTRODUCTION   2

1   DESIGN SOURCES FOR YOUR PROJECTS   4

2   MATERIALS   20

3   BOARD-AND-BATTEN DOOR   36

4   FOUR-PIECE BASEBOARD   46

5   WIDE-PLANK FLOORING   58

6   CHAIR RAIL   72

7   BUILT-IN WINDOW SEAT   82

8   FEDERAL FIREPLACE MANTEL   96

9   DOOR AND WINDOW CASING   110

10   FRAME-AND-PANEL DOOR   122

11   CORNER CUPBOARD   136

12   MAILBOX POST   158

INDEX   168

INTRODUCTION

Everyone loves old houses. Walk through the doorway, and you step into the past: low ceilings, plastered walls, worn floor boards, working fireplaces. From the dry-laid fieldstone foundation to the hand-split shingle roof, these wonderful structures evoke the spirit of days gone by. Simple and utilitarian, yet vigorous and imaginative, they take us back to the early days of America.

Today, many reproduction Colonial houses are being built with commercially available materials, following standard industry practices and current building codes. Builders are pressed to produce cookie-cutter housing quickly to satisfy their customers, the bankers, and their own bottom line, so expensive "frills" are out and "value" is in. The result? Skimpy, pale imitations of an authentic Colonial house. Modern reproductions often lack the proportions and scale, materials and finish, and the robust details of the originals.

If you live in an authentic Colonial house or a modern reproduction and wish to renovate your home in true period style, you may have searched in vain for accurate historical plans or information on building techniques. Every now and then you might come across an article on building a mantel or laying down a wide-plank floor, but these are few and far between. If you will be working on your entire house, what you need is a single unified source to guide you—something that offers authentic plans and details and supplies historical background and information on early tools and techniques, yet tells you how to do the work with modern equipment and materials—and that's what this book can provide.

At first I got interested in period woodwork as a way to showcase my reproduction furniture. But along the way I began to discover fascinating material about the tools and practices of the 18th-century carpenter. I also learned that geography, prevailing customs, and technology had a great effect upon the construction and appearance of a house.

The 10 projects selected for this book present varying degrees of difficulty for the novice and intermediate woodworker. While historical authenticity is a paramount concern of mine, I certainly understand the need to keep the work moving along and not stall out on every obscure detail. You can read about the tools and technology of 18th-century woodworking right alongside modern step-by-step instructions. Each project can be built using common power tools—table saw, bandsaw, router. In most cases, dimension lumber and plywood can be used. This is a realistic approach that places period projects well withing the reach of most woodworkers, both professional and amateur.

1
DESIGN SOURCES
FOR YOUR PROJECTS

A search for authentic period details to incorporate into your house will likely lead you back in time to the buildings of Colonial America. These sturdy and straightforward buildings have much to teach us about decorative woodwork for the home.

Many early American houses grew in stages over the years, as the circumstances of their owners changed. The woodworking inside these houses was affected by technological developments, by changes in styles and taste, and even by the skill of the craftsmen performing the work. For your restoration or reproduction projects to be convincing, they also must seem to have evolved. These stages, considered together, give the house a sense of history and the passage of time. Considered separately, each stage leaves a clear mark.

The evolution of the American farmhouse is a fascinating part of our nation's history. The secret to understanding how these interesting structures arose and developed is twofold. The people who built them played an important role. And certainly as significant were the materials that were available to the builders.

In the late 17th century, most of the settlers in the New England and Virginia colonies were English fishermen, farmers, or shopkeepers. Others were failed tradesmen, runaway apprentices, former soldiers, orphans, vagrants, and even criminals. They weren't wealthy or famous, and they came to America for different reasons: some to make their fortune, others to escape obligations or to gain religious freedom. But they all possessed one thing in common: a sense of adventure and the strength and courage to begin a new life.

Regardless of what propelled them to the New World, they didn't discard their old habits and traditions, their laws, or their sense of order. For the most part, they resumed their English way of life. And that way of life certainly affected the design and construction of their new homes.

EVOLUTION OF THE AMERICAN FARMHOUSE

The first shelters built by the settlers were no more than dugouts, banked into the earth and dug to a depth of 6 ft. or 7 ft., then lined with logs in stockade fashion (see the drawing on p. 6). The roof was laid with split timbers placed directly on top of the walls and covered with moss or sod. Sometimes a floor of sawn boards was laid down. These shelters would last three or four years, until a permanent house could be erected.

As finances and situations permitted, permanent structures began to appear. The housewrights who built these new structures used the methods and materials they had known and used back home. The interiors of these early houses were plain, almost severe. There was little superfluous ornament; everything was there for a purpose. As time passed and the most pressing demand for permanent shelter was met, decorative elements gradually began to appear. Beams were chamfered, paneling was beaded, handrails were molded, and the hearth was dressed with a mantel.

In England, foundations and chimneys were built of brick. Glass was fitted into window sashes. And tiles were used for roofs and occasionally for exterior sheathing. All of these necessary and familiar materials had to be imported from England for use in the Colonies. But importing building materials was expensive and often took years. The need for solid shelter was immediate. So resourceful Colonists had to learn to work with available materials.

If a customary and familiar material was unavailable, a substitute had to be found.

DUGOUT SHELTERS

In the 18th century, farmhouses were built to house generations of a family. If more room was needed, the family simply added on. The photo on the facing page is an example of a sympathetic expansion of an old house.

Unlike the stud frames common in homes of today, the frame of a typical house built in Colonial times was composed of massive oak beams that were pegged together (see the drawing on the facing page). In these timber-frame buildings, beams served as the skeleton of the structure and defined the shape or outline of the house while supporting the walls, floors, and roof. These massive members were cut to fit into one another (see the sidebar on pp. 18-19) and remain firmly attached. Each brace and floor joist added to the rigidity and strength of the completed structure.

The timbers for an early house were always hewn, even after sawmills were in widespread use. Why? These timbers were so large that hauling them to a sawmill, mounting them on a saw carriage, squaring them, and hauling them back to the building site would have been far more difficult than simply felling the trees on the site and shaping them in place with a hewing ax.

Fieldstone replaced brick, oiled paper replaced glass in windows, and wood was used for almost everything else. Wood was the perfect substitute for any unavailable building material. This country was literally covered with trees—it was one gigantic forest. In order to build a house or a road or plant crops, trees had to be cut down to clear the way. Timber was everywhere, and it was versatile, easy to work, durable, attractive, and familiar to the settlers. Oak was used for house frames, cedar for roof shingles, and pine for interior millwork, such as doors and floor boards.

Each wing of this 200-year-old farm-house in upstate New York has maintained the scale and the spirit of the original structure.

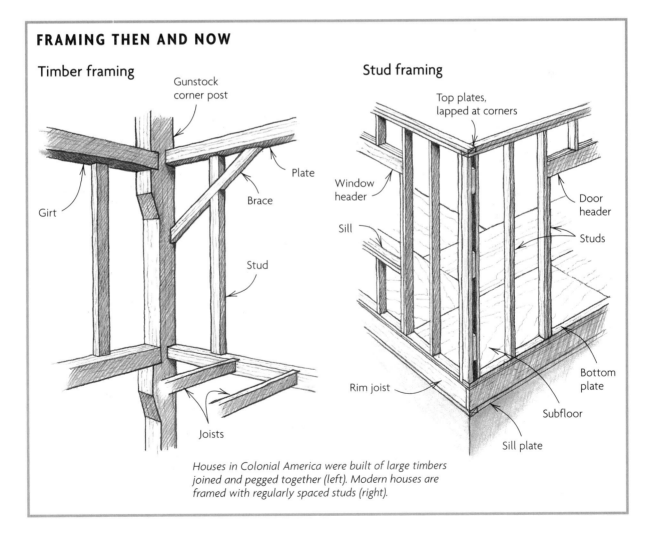

FRAMING THEN AND NOW

Timber framing

Gunstock corner post

Plate

Brace

Girt

Stud

Joists

Stud framing

Top plates, lapped at corners

Window header

Door header

Sill

Studs

Rim joist

Bottom plate

Subfloor

Sill plate

Houses in Colonial America were built of large timbers joined and pegged together (left). Modern houses are framed with regularly spaced studs (right).

Some elements of early American woodwork, such as the frame-and-panel construction of the doors on this modern built-in bookcase, have never gone out of style.

By the mid-19th century, timber framing gave way to stick framing and stud-wall construction, the standard method of building today.

INTERIOR WOODWORK

After a house was framed, roofed, and sided, attention turned from basic structural work to interior woodwork, which was often a showcase for the carpenters' creativity and craftsmanship. Rooms were entered through handsome doors, windows were framed by elegant chair rails, and the junction between walls and floors was dressed by tall baseboards. Other examples of fine woodwork abound. Fireplace hearths were considered naked without a proper mantel, and a corner cupboard could transform a simple room into an elegant parlor.

One of the most common construction methods used in Colonial interiors was frame-and-panel construction. Within a frame, a solid wood panel was held in position as part of the design, but was free to shrink or expand with changes in humidity. Frame-and-panel construction permitted large areas to be covered with wood without the risk of cracks and splits. It was used in the construction of wall paneling, doors, cabinet sides, dust panels, cabinet backs, and shutters. This type of construction has changed little over the years (see the photo above), and I've used it extensively in many of the projects in this book. See pp. 18-19 for other joinery methods for interior woodwork.

DEVELOPING YOUR DESIGN

As you design your project, you must be sensitive to the nuances of architectural history in your own house if it is old or in the houses you're using as models. There are several good ways to increase your knowledge of architectural detail. Historic houses and historical societies have much to teach, as do agencies of the federal government, such as the Historic American Buildings Survey (HABS). If you live in an old house, you can do a little detective work of your own (see pp. 14-17).

Historic restorations

One of the most enjoyable ways to develop a feel for early architecture is to visit historic restorations. These may be part of museums or entire villages, such as Colonial Williamsburg in Virginia or Sturbridge Village, Historic Deerfield, and the Hancock Shaker Village in Massachusetts. When you go, plan to spend a day, if you can, and walk through and study as many historic examples as possible. Many of these places have costumed interpreters, and most have collections and research libraries.

In short order, you will realize that houses in the 18th century were not the sleek and conve-

Houses in the 18th century provided shelter and warmth, but not much in the way of amenities. This log cabin at Museum Village in Monroe, New York, has only three windows.

In early farmhouses, a room had to serve several purposes. This 12-ft. by 14-ft. room at Museum Village in Monroe, New York, functioned as a kitchen, dining room, bedroom, and workroom.

nient palaces we crave today. Houses in the 18th century didn't have bathrooms, and they didn't have much closet space. Their windows were small and their doors narrow. Old houses were often the hub of the working farm; they sheltered a large family and possibly hired hands. They provided storage for food and tools, and they functioned as laundry and kitchen.

In these houses, the best woodwork is found in the parlor and front rooms. The fanciest mantels, the best moldings, and the largest windows were put where they would be seen by guests. Back rooms had board-and-batten doors instead of six-panel doors, simple balusters instead of carved ones, plain vertical paneling instead of full paneling. This practice was common in the 18th and 19th centuries and is a very effective way to evoke a period effect.

Walking through a period restoration helps you to develop a sense of scale. I know a contractor who built a meticulous Cape Cod reproduction house.

All the architectural detail was there—molded corner columns, clapboard siding, and period-style lanterns flanking the front door. But when entering the house, you immediately noticed the nearly 9-ft. ceilings, the wide doorways, and the oversized windows. The scale of the house was all wrong for the period.

Historical societies

Historical societies are a great source of information about period details. These groups have something of a bad reputation as obstructionists—I'm sure you've

Much can be learned from local historical societies. Here the author, left, and Michael Bertolini, curator of the Warwick Historical Society, examine an early Federal fireplace mantel at the Azariah Ketchum house in the historic district of Warwick, New York.

heard stories about meddlesome reactionaries who prevent homeowners from tearing down decrepit outbuildings, building additions, or even painting the exterior of their home, They cast local historical groups as busybodies whose only concern is maintaining the value of their properties.

Such occurrences are often widely publicized and blown out of proportion. Historical societies are more likely to be composed of local citizens deeply interested in history, social customs, and the preservation of early architecture. In Warwick, New York, the local historical society strives to assist the old-house owner or

builder. Architects, builders, interior designers, lawyers, and teachers volunteer their time to monitor and advise homeowners in their renovations. This local group is not part of the government, and so has no authority to deny building permits or stop construction, except in designated historic districts. But their advice can often prevent the new homeowner from making costly mistakes and incurring long construction delays.

Whenever I've needed the services of a contractor sensitive to an old house—a blacksmith, stonemason, or even an upholsterer—I have gotten invaluable help from the historical society. There is even a historian who helps research the history of local homes. One group I work with sponsors social events and fundraisers to pay for the upkeep and improvement of several historic properties belonging to the society. It also opens the houses to the public for tours. This group even offers loans below commercial rates to help homeowners and commercial businesses improve their properties.

Historic American Buildings Survey

In 1989, while browsing in a bookstore that specialized in art and architecture books, I opened up an old volume that contained an exquisite drawing of an 18th-century farmhouse. Intrigued, I took a closer look. It was an old catalog of the Historic American Buildings Survey (HABS). The book's directory included listings of almost 28,000 historic houses, organized by state and town. From that old catalog I was able to obtain enough information to

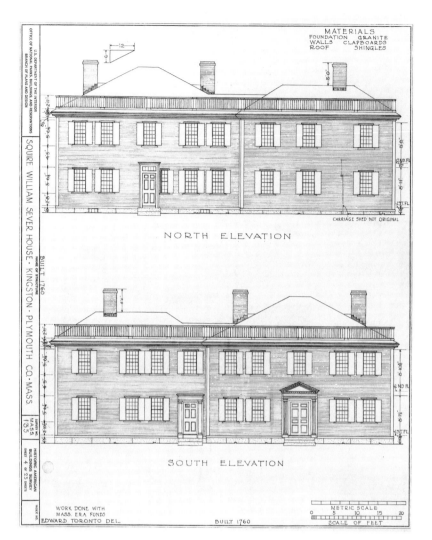

Two typical drawings from the Historic American Buildings Survey show the wealth of detail available to woodworkers who are willing to do a little research.

track down a recent edition that explained how to order prints and photos. I was stunned that such a marvelous resource was available and accessible to anyone. I had never heard of it before.

During the 1930s and 1940s, the federal government funded numerous public-works pro-

grams to get the country back on its feet after the Great Depression. Many of these were building projects, such as roadways, public parks, airports, and dams; there were a number of artistic projects, such as murals and sculptures, as well. Sandwiched in between was the Historic American Buildings

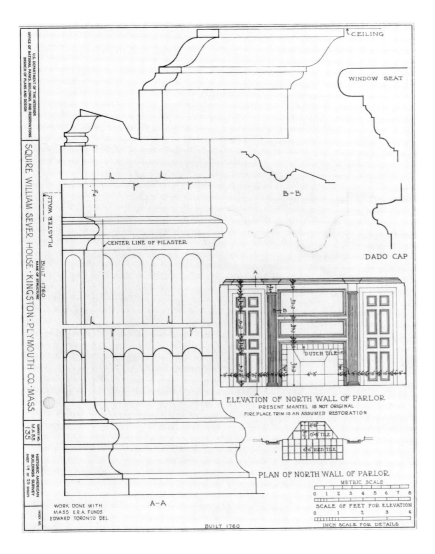

U.S. DEPARTMENT OF THE INTERIOR
OFFICE OF NATIONAL PARKS, BUILDINGS, AND RESERVATIONS
BRANCH OF PLANS AND DESIGN

SQUIRE WILLIAM SEVER HOUSE · KINGSTON · PLYMOUTH CO · MASS.
NAME OF STRUCTURE
BUILT 1760

SURVEY NO.
MASS.
135

HISTORIC AMERICAN BUILDINGS SURVEY
SHEET 14 OF 25 SHEETS

INDEX NO.

CEILING

WINDOW SEAT

B - B

DADO CAP

PLASTER WALL

CENTER LINE OF PILASTER

A - A

WORK DONE WITH
MASS. E.R.A. FUNDS
EDWARD TORONTO DEL.

BUILT 1760

ELEVATION OF NORTH WALL OF PARLOR
PRESENT MANTEL IS NOT ORIGINAL
FIRE PLACE TRIM IS AN ASSUMED RESTORATION

DUTCH TILE

6-6 RED TILE

PLAN OF NORTH WALL OF PARLOR

METRIC SCALE
0 1 2 3 4 5 6 7 8

SCALE OF FEET FOR ELEVATION
0 1 2 3 4

INCH SCALE FOR DETAILS

and completeness. Some of the buildings are recorded in elevations, floor plans, and sections, with hundreds of details—molding profiles, sketches of wrought-iron hardware, even the placement of rosehead nails on board-and-batten doors. These drawings contain all the information needed to reconstruct some of this country's minor architectural masterpieces.

It's important to note that the drawings show the woodwork and architectural details as the viewer would see them; there are few cross sections revealing the construction of a particular element. Anyone who wanted to reproduce a doorway or mantel would need to be familiar with the construction of other period examples to make an educated guess as to its construction. Still, the information contained in the drawings is invaluable and can shed much light on the evolution of domestic architecture and building practices as well as on the small-scale trades and social customs of the day.

The Library of Congress has published a guide, called *America Preserved: A Checklist of Historic Buildings, Structures, and Sites* (1995), that lists more than 30,000 officially designated historic structures that are documented by HABS. In this hefty 1,184-page book, which at this

Survey. The purpose of this project was to employ architects, engineers, surveyors, and photographers to record historic American buildings of important design and in original condition. Squads were dispatched all over the country to record data, take photographs, and render exact drawings of selected early structures.

In the last 50 years the value of this endeavor has become apparent. Many of the buildings drawn and recorded have since succumbed to the wrecker's ball in the name of progress or to fire and other natural disasters. But their images and details are preserved forever. The drawings that were made (see above and the facing page for two examples) are remarkable for their detail writing sells for $37, the structures are listed by state, county, and town. There is no index or cross reference by architectural style or date. The capsule descriptions are very brief, but they do tell you what drawings and photos are available. Here is an example: "New York State, Onondaga County, Syracuse, General Leavenworth House, 607 James St. Wood, 2½ stories, mid 19th century, mansion type, Greek Revival, 9 sheets [of drawings] (1934); 3 photos (1934)." The Library of Congress can provide a researcher to check up to five buildings or structures for the information you seek. This service is free, but the request must be made in writing to the Library of Congress Prints and Photographs Division, James Madison Building, Room 337, Washington, DC 20540-4730. The LOC home page is http://lcweb.loc.gov.

The National Park Service also maintains a HABS web site (http://www.cr.nps.gov/habshaer/habshome.htm) that provides detailed information about each structure. This information can help you zero in on what you need. There are no drawings on line at the present time, but graphic files are expected to be available soon. When they are, you'll be able to print out a drawing in an 8½-in.-by-11-in. format and decide if it looks promising. Then you can order full-sized (18-in.-by-24-in.) prints from the Library of Congress Prints and Photographs Division. All you have to do is identify the structure by name, city, county, and state. The prints can ordered in one of two ways: as photocopies or as semi-transparent architectural drawings (Diazo prints). The photocopies cost 50 cents a sheet, and often these are clear enough to yield the information you need. The Diazo prints cost about $6 per page.

TRACING THE HISTORY OF AN OLD HOUSE

If you live in an old house, you may wonder what it looked like a century or two ago and how it came to look as it does today. And if you are looking to renovate that house, knowing its history can help you immeasurably in developing your designs.

My own house is a three-story rambling structure, set on 2½ wooded acres at the end of a country road in Warwick, New York. I've lived here for the last six years, and I decided to trace its history, with an eye to understanding how it grew into a comfortable home for a modern family. The height and girth of the structure have tripled over the years, but the bones of the original cottage can still be recognized, if you look carefully. Because many of the projects in this book were designed for my own house, I'd like to share my discoveries with you, in hopes that you will undertake similar research on your own house.

I began by studying other homes in the area, noting the size and number of windows, the molding profiles, the width of the floor boards, and the types of doors and shutters. As I examined the stone and brickwork, the surfaces and dimensions of the framing members, and even the types of nails and fasteners used, the age and chronological order of alterations and improvements became clear. I also went to the town hall for property records and enlisted the aid of the local historical society. I learned quite a bit. You can also follow the development of the house in the drawing sequence that begins on the facing page and continues on p. 16.

Around 1780, Abraham Van Dalson settled on a large tract just north of the village of Warwick. In 1790, his son John began work on a simple one-room cottage, roughly 20 ft. square. The timber frame of the house followed the traditional English prototype. This was a

EVOLUTION OF A FARMHOUSE

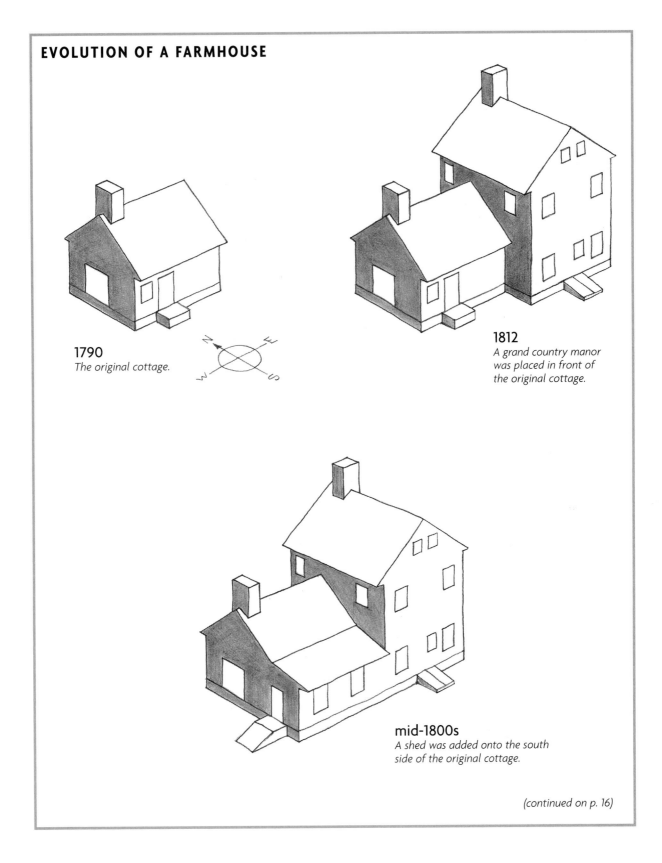

1790
The original cottage.

1812
A grand country manor was placed in front of the original cottage.

mid-1800s
A shed was added onto the south side of the original cottage.

(continued on p. 16)

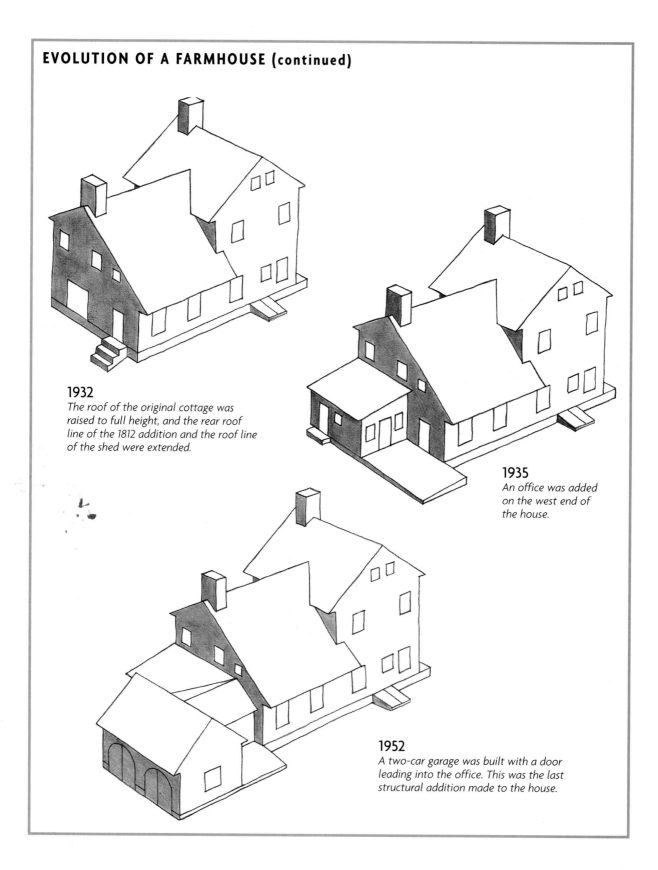

1932
The roof of the original cottage was raised to full height, and the rear roof line of the 1812 addition and the roof line of the shed were extended.

1935
An office was added on the west end of the house.

1952
A two-car garage was built with a door leading into the office. This was the last structural addition made to the house.

small house with a cramped sleeping loft and a low ceiling; it served as kitchen, dining room, laundry, and bedroom for up to six adults and children.

Farming was a hard life, but the younger Van Dalson must have been successful at it, because in 1812 he was able to build a rather large and fancy "addition" to his small cottage. (Actually, it was more like a large new house added to the old one.) This two-story structure, built in 1812, was by any standard a luxurious manor house. A house of this size (20 ft. by about 35 ft.) usually had its door and hall centrally placed, dividing the floor plan into two rooms on each floor. In the Dolson house, however (John had by this time changed his last name), the door was at one end, opening into one large room with a spacious entry and hall on the ground floor. This was an unusual plan that, combined with high ceilings, elaborate woodwork and fireplaces, created a sense of light, luxury and spaciousness.

When the new structure was completed, the original cottage, which contained the large cooking fireplace and baking oven, became the kitchen and dining room. The loft overhead remained a bedroom for children or guests. The new house was situated in front of the old cottage, closer to the road, so

passers-by were presented with its grand facade; the old cottage was hidden from view.

Sometime between 1812 and 1870, a small shed was built onto the south side of the old part of the house. It might have served as a pantry, or as a tool shed; it might even have been a chicken coop.

In 1932, the house and 26 acres of land were bought by a land surveyor named Pierson Booth and his wife, Helen. At first they lived in the farmhouse as they found it. After a couple of years and with a baby on the way, it was time to add on and to modernize.

The Booths decided to install a coal-fired boiler, an indoor bathroom, and a dedicated kitchen. Either they themselves or their builder came up with an ingenious plan for the expansion. Instead of building another addition, which would have required new foundations and spread the house out, they decided to add space vertically. The original cottage and the lean-to were raised to two full stories. Then the rear roofline of the 1812 addition was extended to meet the new height of the original cottage and the lean-to. These additions, unlike the rest of the house, were stick-built (framed with 2x4s) instead of timber framed.

In 1935 a fire broke out in the old kitchen chimney and spread through the overhead space above the old sleeping loft, finally breaking through the roof and into the back hall. Luckily, the fire was put out quickly and the damage was not severe. Many of the charred studs and rafters are still in place, with new studs and rafters scabbed onto them. At the same time, a lean-to was added on the west end of the house to serve as an office. Here Booth kept an oversized drawing board and his surveying tools. He heated the office with a small potbellied stove that now warms my woodworking shop. You might be wondering how I discovered the date of the fire. Elementary—the workers who built the office signed and dated their work on the sheathing under the exterior shingles.

The most recent addition came in 1952 (the date was inscribed into a foundation footing), when Pierson Booth built a two-car garage onto the house with a door on the north side of the house leading to his office. The garage was stick-built and covered with cedar shingles to match the house. At this time, new windows were installed in the bedroom over the old kitchen, in the office, and in the garage.

JOINTS FOR TRADITIONAL WOODWORK

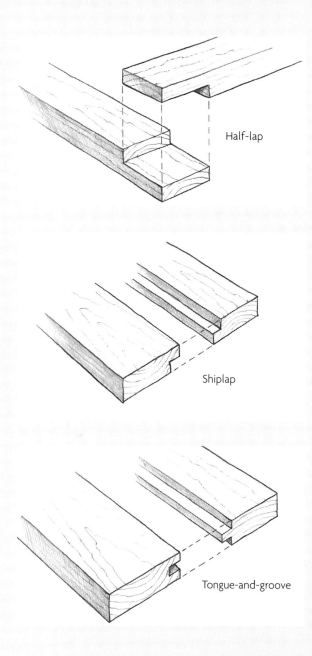

Half-lap

Shiplap

Tongue-and-groove

The joints used in timber-frame structures and early interior woodwork are still in use today. These include the half-lap, the shiplap joint, the tongue-and-groove, the mortise-and-tenon, the dovetail, the dado, the groove, and the rabbet.

HALF-LAP

In the half-lap, half of the material's thickness is removed from each member across the joint. When the parts are lapped, one over the other, they form a joint the full thickness of the members. In a timber-frame house, the sill and the rafters are half-lapped. For an unusual variation on the half-lap joint, see the door and window casing in Chapter 9.

SHIPLAP

The shiplap is similar to the half-lap, but it is cut along the edges of the boards, not at the ends. The shiplap is a joint that accommodates the shrinkage and expansion of wood without showing a gap. In Colonial America, this joint was most often used for flooring (see Chapter 5) and in vertical and horizontal paneling.

TONGUE AND GROOVE

The tongue-and-groove joint is the most effective way to join boards on edge. Like the shiplap, it allows for wood movement across the grain. Also, since the tongue on one board fits into a groove on the mating board, the surface alignment of the members is preserved and the resulting panel is flat. Tongue-and-groove joints were commonly used for flooring and wainscot paneling, as well as for tabletops. In this book, the board-and-batten-door project (Chapter 3) uses tongue-and-groove joinery.

MORTISE-AND-TENON

The mortise-and-tenon joint was used extensively in the construction of doors, mantels, and paneling. This is the best method for joining two pieces of wood at right angles on edge because if offers the most long-grain to long-grain contact. The strongest version of this joint is the through haunched mortise-and-tenon, used in the frame-and-panel-door project (see Chapter 10).

DOVETAILS

Dovetailing is the strongest way to join two pieces at right angles. The dovetail is one of the most beautiful joints in woodworking, and one of the most difficult to execute. Because of the flared tails, the joint is impossible to pull apart in one direction. Dovetails were used to join cabinet carcases, blanket boxes, and drawers.

DADOES AND GROOVES

Dadoes and grooves are cut through the middle of a board. The dado is cut across the grain of one board to receive the thickness of another at 90°. Its most common application was to join shelves and partitions into case sides. The groove (also known as plow) is cut along the grain. It was commonly made around the inside of a frame to receive a panel.

RABBET

The rabbet is a step cut along the edge or end of a board to receive the thickness of another board at 90°. It was a simple cut to make and was widely employed in Colonial America for setting backboards on cabinets. It was also used to step down stock for moldings (see the sidebar on pp. 55-57).

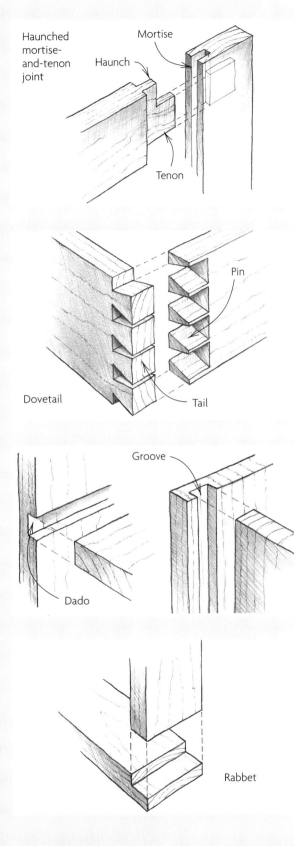

2 MATERIALS

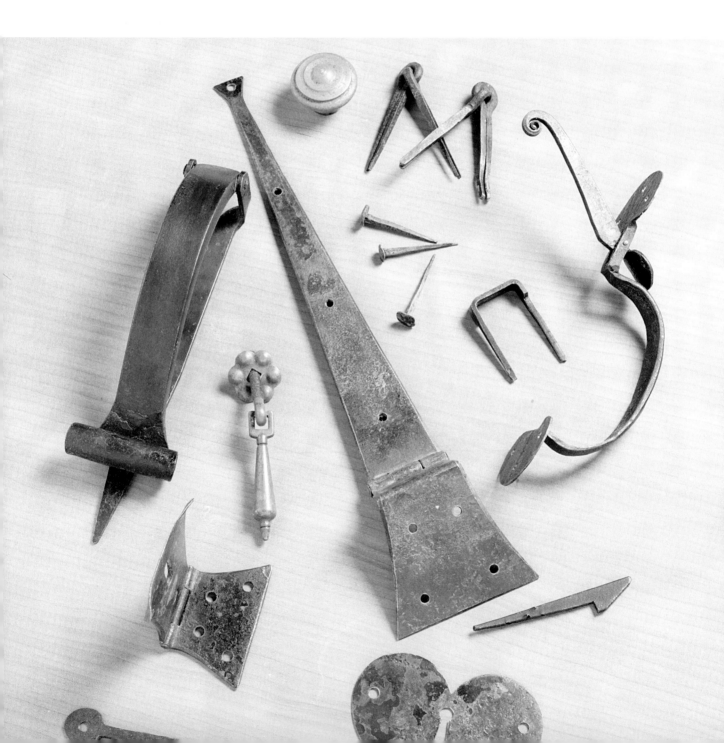

In addition to researching architectural details for your project, you'll need to take a look at materials, both traditional and modern. These materials will include wood, hardware, and finishes. What you choose will depend on how authentic you want your project to be. Is it a restoration, where everything chosen should replicate original materials as closely as possible? Is it a remodel, where little attention is paid to historical authenticity? Or is it a renovation, where modern materials can be used to replicate a period look? To help you with your decisions, we'll examine both the traditional materials and their modern equivalents.

TRADITIONAL BUILDING MATERIALS

Historically, pine was the choice of the Colonial woodworker. In the Northeast, pine trees were plentiful and grew to enormous size, yielding straight, clear boards 2 ft. or more across. (In my own house, I pulled up quite a few 20-in.-wide floor boards.)

When a tree was sawn into boards, it produced only a small percentage of quartersawn material (see the drawing below). Quartersawn boards (with the rings more or less vertical across the end grain) was prized for its dimensional stability and its

good looks. Colonial woodworkers saved this choice material for critical door, cabinet, and furniture parts. It was especially good for moldings. Since a molding plane can be passed only from right to left in the direction of the grain, straight, clear wood had to be used in order to obtain a full, clear profile with no tearout.

The remaining plainsawn stock was used for panels, flooring, shelving, and other unseen or lesser work. Boards with loose knots, wany edges, and other defects were used for exterior sheathing, subflooring, and attic floor boards.

PLAINSAWN VS. QUARTERSAWN STOCK

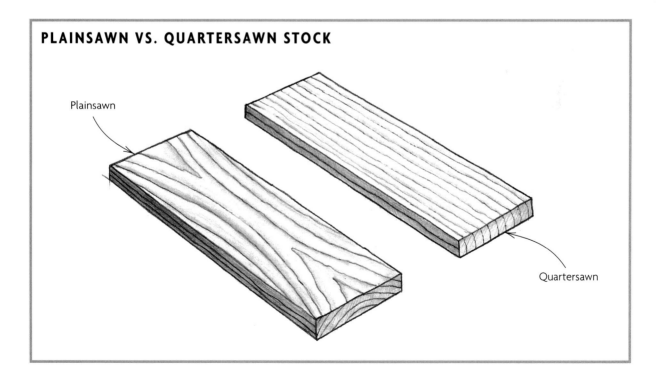

Plainsawn

Quartersawn

MODERN BUILDING MATERIALS

Pine is still the ideal wood for traditional interior projects, whether it will be painted, stained, or left its natural color. It has a pleasing grain and mellows to a soft, rusty hue after a few years.

Pine is also easy on hand tools, sawblades, and router bits, allowing work to progress quickly with excellent results. And since pine is so soft, sanding goes quickly and sandpaper lasts longer. Pine also takes paint well. If the wood is properly prepared with primer, sanded, and filled (see pp. 29-31), paint will coat the surface nicely without rough spots or telegraphing grain.

In the last few years the price of clear pine has skyrocketed, and in my area it's almost as expensive as mahogany. Still, for the sake of authenticity, I choose pine for all my interior woodwork. In order to save money, I buy #2 common pine from the local lumberyard material at a fraction of the cost of clear lumber. I go through the stacks one board at a time, looking for the best boards. Most of the boards have small scattered knots, but will still contain a surprising amount of clear and straight material. Depending on how the lumber is stored at the yard, its

moisture content may not be ideal, so to be safe, I sticker and stack it in my shop for at least three weeks before using it. This method has been so successful that I rarely buy clear pine any more.

Plywood
Colonial woodworkers had to assemble large panels by gluing up narrow boards; modern woodworkers can simply use a piece of plywood. Plywood offers a number of advantages over solid wood. It comes in 4x8 sheets, so you just cut a panel the size you need: There are no edges to joint, glue up, and fair. Because of its cross-grain lamination, plywood is stable and strong. For paint-grade woodwork, plywood provides a smooth and uniform wood surface that leaves a beautiful paint finish and won't telegraph seams later. I recommend using it wherever possible. All the cabinet projects in this book employ plywood where it will save time and money without compromising design.

Medium-density fiberboard (MDF)
Like plywood, medium-density fiberboard comes in 4x8 sheets. It is a manufactured wood product made of of pulverized wood fibers that have been heated and compressed after resins have been added to improve strength and stability. MDF is cheaper

than plywood. Its surface is ideal for paint, and it isn't affected by changes in temperature or humidity. However, MDF doesn't hold fasteners well, it's hard on blades and cutters, and it sags under its own weight. In spite of these drawbacks, there are still applications where MDF is unsurpassed. Since it is a paper product, it has no end grain, so the customary and common problems with end-grain milling and finishing are avoided. I've used MDF for raised panels that were to be painted and got great results. MDF also works well in vertical applications, such as partitions within a cabinet.

TRADITIONAL HARDWARE

In the 18th century, nearly every village had a blacksmith who made everything from nails and cooking utensils to farm tools and architectural hardware. Early forged-iron work was utilitarian in design and appearance, but later work was embellished with swags, rattails, and cusps. These catches, latches, hinges, locks, and straps often surpassed in beauty the woodwork to which they were attached. Forged-iron hardware is much appreciated today. There is a lively market for it and a growing one for high-quality reproductions.

Because of the low-carbon composition of the metal and its inherent impurities, wrought iron has a mottled surface texture. Since early hardware was hand wrought, or shaped by hand, it was often irregular in thickness and in outline. These qualities give early hardware real character and a unique appearance that complements period woodwork.

Architectural hardware was treated by the blacksmith with a forge finish: a kind of shiny crust that thwarted rust. I have a number of old pulls and hinges in my house that were never painted or protected in any way, yet they are rust free today.

Since interior doors were less than 1 in. thick, early hardware was often surface mounted and completely visible. Knowing the hardware would be on display, craftsmen took the opportunity to integrate their hardware into the design of the woodwork and make it an exciting and beautiful part of the room.

In the 18th and 19th centuries, blacksmiths made all sorts of hinges, latches, pins, and hasps for securing doors and lids. Most of them followed English patterns, but there were endless regional variations on the designs.

Snipe hinges

The simplest hinges, called snipe hinges, were used mostly on blanket-box lids and small cupboard doors. Snipe hinges were made from slender wrought-iron rod. These were two cotter-pin-like pieces connected together. One end was driven into the underside of the lid and the other was set into the back of the box or the cabinet stile. The protruding ends were then clinched over. Snipe hinges rarely worked smoothly and had to be operated carefully.

Strap hinges

Strap hinges from 8 in. to 24 in. long were the best way to support large, heavy doors. The end of the strap hinge that tapers in width and in thickness was nailed to the surface of the door; the other end, with an eye that swings on a pintle, was driven into the door jamb (see the photo below).

Butterfly hinges

Like most early hardware, the butterfly hinge was surface-mounted to the door and the cabinet stile. Its two leaves flare toward the ends, resembling a butterfly. The knuckle of the hinge is a three-piece barrel, held together with a straight pin (see the photo at left on p. 25).

A sturdy strap hinge can support a heavy entrance door. The eye of the strap is mounted on a pintle, which is driven into the door frame.

A thumb latch, seen from the exterior (right) and the interior (below) sides of the door. Depressing the flat on the latch bar lifts the curved end on the other side, which in turn raises the pivoting bar enough to clear the hook so the door can be opened.

H and H-L hinges

Because of their narrow, but long, leaves, H hinges (see the photo at right on p. 35) and H-L hinges were used instead of strap hinges on large interior doors. Since their barrel is longer, they provided good support for a door and required less clearance than a strap hinge mounted on a pintle.

Thumb latches

Before locks were used in the Colonies, doors were secured with a thumb latch. On the exterior side of the door, a forged handle was anchored at each end by a cusp (see the photo at top left). The upper cusp was pierced by a latch bar, which had a flat on one end and a downward curve on the other. On the interior side of the door, a pivoting bar was secured by a hook driven into the door casing (see the photo at bottom left). To enter the house you depressed the latch bar, which raised the other end and the pivoting bar along with it, clearing the hook and allowing the door to open. To leave the house you simply raised the curved end of the latch bar.

This hand-forged butterfly hinge has a beautifully mottled surface. It is mounted to the woodwork with rosehead nails that are clinched (bent over) over to prevent loosening.

You can make imitation rosehead nails by heating cut nails and flattening their heads.

Thumb latches really help to create a period atmosphere in a room. And like the crackle of logs burning on the hearth or the creak of old floor boards, the distinct sound a thumb latch makes adds another dimension to a Colonial interior.

Knobs and handles

Early metal knobs were cast of brass in three parts, then silver soldered together. A threaded iron stem was fitted to the end. Other knobs were made of wood and turned on a lathe. In terms of design, the early knobs are not very different from those made today.

Fasteners

There are several ways to determine the age of an old house, but one of the surest is to examine the nails and hardware. During the early 1700s, nails were made exclusively by hand. Ken Schwarz, an interpreter-blacksmith at Colonial Williamsburg, told me that 18th-century blacksmiths typically produced about 800 hand-forged nails in one day, Nailers (smiths who specialized in making nails) could forge 2,300 nails in the same period.

It wasn't until the late 1700s and early 1800s that nailmaking machines were perfected. These machines produced cut nails— flat tapered nails easily distinguished from forged nails, which had four tapering sides and a distinctive faceted head.

Although a good reproduction will require the proper hand-made fastener, hand-forged nails are prohibitively expensive. That's why I often reheat cut nails, then reform and flatten the head to resemble early nails (see the photo above right).

MODERN HARDWARE

You needn't limit yourself to reproduction hand-made hinges, locks, and catches for your traditional woodwork. That's a good thing, because antique hardware isn't always available and hand-made reproductions can be very costly. Fortunately, there's a way to make new hardware look old (see the sidebar on p. 35).

There is an endless array of modern hardware to choose from. Many items are designed for fast and easy installation, maximum adjustability, and broad application. A number of hardware catalogs offer high-quality hardware for almost any design situation.

Butt hinges

The most common modern hinge is the butt hinge. Each leaf of the butt hinge has a crenelated edge that interlocks with the other leaf to form the hinge barrel; a hinge pin holds the leaves together. Butt hinges are used on inset-door installations when a clean, traditional appearance is desired. The piano hinge, a variation on the butt hinge, is simply a continuous butt hinge used to support large lids on window seats or accordion doors. I don't use piano hinges because I don't like the way they look and I find them tedious to install.

Euro-style hinges (32mm)

Euro-style hinges have become the most popular hardware for doors on built-in and kitchen cabinetry. These hinges, also called cup hinges, are easy to install. They are available in a variety of designs for a range of applications and are visible only when the cabinet is opened. What I like best about these hinges is they can be adjusted with a screwdriver to accommodate an unsquare cabinet or installation site. They are available for both inset and overlay applications.

Another popular form of the Euro-style hinge is the pocket-door hinge. This hinge is mounted on a sliding track. When the door is opened to 90°, it can be slid back into the cabinet with only a few inches protruding. Pocket-door systems are perfect for hiding televisions and audio equipment in traditional built-in cabinetry.

Catches

Magnet catches have a small metal plate that is attached to the inside corner of the door. When the door is closed, this metal plate is attracted to and held in place by a magnetic base piece. Magnet catches are effective, but unsightly.

Bullet catches are compact barrels with an adjustable spring-loaded bullet-shaped head. The head protrudes from the edge of the door slightly and snaps into a small pan that is set into the door opening. Unlike magnetic catches, bullet catches are discreet and hardly noticed.

Locks

Full-mortise and flush-mounted locks are effective ways to secure a door or drawer and add a period touch. These can be difficult to install, but I think they're worth the effort.

Drawer slides

Since they won't be seen when the drawers are closed, modern ball-bearing drawer slides are a great way to ensure smooth and quiet operation of drawers, good load-bearing capacity, and simplified cabinet design.

Fasteners

Fasteners have come a long way since the days of cut nails. The modern woodworker can select from a vast array of machine-driven screws, staples, and nails.

Screws I use screws to join practically all of my built-in casework. They are strong, fast, rustproof, and easy to install. They are also easy to remove, should you make a mistake. It's best to keep on hand a variety of sizes and lengths. My favorite is the galvanized Phillips-head sheet-metal screw. In my work I regularly use #8s and #10s in the following lengths: 1 in., 1¼ in., 1½ in., 1¾ in., 2 in., 2½ in., and 3 in.

For the best results, I recommend predrilling and countersinking each screw, then driving them with a heavy-duty drill or driver. The same tool can also be used to take out any misdirected screws. Errant screw holes can be filled with putty or plugged with wood.

Air-powered staples and nails
Today almost every professional builder uses pneumatic nail guns. One squeeze of the trigger shoots and countersinks the fastener in less than a second, and you can position the workpiece with one hand while firing the nail gun with the other. The best thing about using nail guns is a blemish-free job—there are no split moldings or hammer marks to repair or disguise later. You might hesitate to outfit your shop with pneumatic equipment because of the initial cost, but this is one expense you'll never regret.

There are three different pneumatic fastening tools I find useful: a mini-pinner, a finish nailer, and a narrow-crown stapler. The mini-pinner is a small gun that shoots tiny 21-gauge headless pins (see the photo above right).

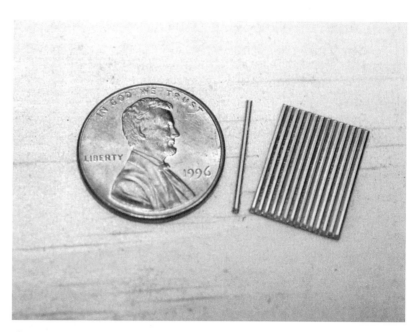

The 21-gauge pins fired by the mini-pinner leave holes so tiny they can be filled completely with a small dab of wax.

Because of its size, the pinner has no recoil, and so it doesn't dimple the wood surface. It leaves a hole so small it can be filled with a quick smear of wax. A mini-pinner is the perfect tool to use for installing thin moldings or small parts, and it can be used to strengthen outside corner miters. The finish nailer fires small 18-gauge nails that range in length from ⅝ in. to 1⅝ in. This gun is suitable for attaching large moldings, glue blocks, and cockbeading. The narrow-crown stapler shoots staples from ½ in. to 1½ in. long ("narrow crown" refers to the width of the staple and the narrow entry hole it makes). These staples have good holding power because the prongs converge in the material.

The tool is great for joining plywood cabinet parts, attaching cleats and cabinet backs, and assembling drawers.

TRADITIONAL FINISHES

If you examine the walls and woodwork of old houses, you'll see different types of finishes, depending on the age of the house and the economic status and taste of the owners.

Wallpaper
If you have ever stripped wallpaper in an old house, you have probably been delighted to discover layer upon layer of old

material that hints at the styles and tastes of bygone ages. Most early wallpapers came from England. Later, the most popular designs were imported from France. American wallpaper designs with simple patterns and limited colors began to appear by the mid 18th century. Contrary to what most people think, wallpaper was wildly popular and by the 1830s it had replaced paint and paneling as the wall covering of choice.

As patterns and colors became more sophisticated, wallpaper became an important design element in the decoration of a room. As a result, architectural features such as fireplace mantels, moldings, and other woodwork were painted in lighter, softer colors that would not compete with the vibrant patterns on the wall.

Paint

In the 17th century, paint was expensive and hard to come by. It was not uncommon for the exterior of houses and other woodwork to be left bare without the protection of paint. By the early 1700s, plaster walls were either left bare or whitewashed and the woodwork was painted, generally in earth colors. In the 1730s, Prussian blue was popular, but very expensive and usually reserved for the wealthy.

By the late 1700s, wallpaper had replaced wood paneling in most rooms, and architectural moldings and details were painted in lighter colors, allowing the wallpaper to become a prominent element in the decorative scheme.

By the early 1800s, paint was affordable, and readily available in colors that ranged from bright primaries to soft pastels. During this time, plaster walls were painted a light color with the woodwork painted a darker color.

In the early 1700s, paints were not manufactured, and professional interior painters had to mix their own by combining oil with natural pigments obtained from plants, minerals, or clay. The pigments that were available at the time produced a small range of colors—red, brown, taupe, mustard, and yellow. Milk paint was one such homemade concoction. Skimmed milk, whiting (chalk), lime, and boiled linseed oil were mixed together, then colored with pigments such as umbers, siennas, Venetian red, or yellow ocher.

The texture of early paint was terrible—thick and slightly lumpy. It didn't spread well or self-level like modern paints and had to be worked into the wood with coarse brushes. This left a striated and uneven surface.

Since early paint was hand made, the colors weren't homogenous. Paint was prepared in small batches, so obtaining a match with the next batch was a bit of a trick. When several pigments were combined to create a new color, spots of the individual pigments would often appear, giving the painted surface a mottled, almost speckled look.

Durability and lightfastness were also problems. Early paints often faded when exposed to sunlight. If a painted surface was washed, the paint was often removed along with the dirt. Sometimes in an effort to protect the paint, prolong its life, and give it a light gloss, the woodwork was coated with linseed oil. But the oil frequently darkened the paint, altering the original color.

The instability of early paint and the effects of time on a painted surface make matching an early painted surface difficult at best.

MODERN FINISHES

Almost every paint company offers an Early American collection of special shades, inspired by historic museums or Colonial Williamsburg, that are considered to be authentic. The same holds true for wallpaper. Most paint stores have sample books devoted to 18th- and 19th-century patterns.

Selecting the right finishes for your projects is important for a convincing and realistic effect, but don't get carried away with your zeal for authenticity. In the past, I've had many clients who insisted that I finish their project with authentic milk paint. I always asked them if they really knew what milk paint looked like. Insulted by my question, they would reply, "Of course!" Inevitably, when I delivered the furniture, they would complain that the paint looked splotchy and uneven, with wood grain showing through in spots. The fact that I used authentic milk paint proved to be no defense— I usually wound up repainting the furniture.

Eventually I got wise. I figured out what they really wanted: a soft, dusty, almost matte surface, buffed to a soft gloss where people touch it. They wanted bare wood showing at the edges and a light paint buildup in the corners, but without drips. They expected soft colors that were easy on the eyes and suggested age, not the harsh limited palette of old-fashioned milk paints. What they really wanted was 18th-century charm with 20th-century perfection. My solution to this was modern latex paints.

Latex paints have many advantages over traditional paints. Since such an incredible range of color is available, almost any shade can be duplicated at reasonable cost. Today, in addition to carrying extensive color palettes and custom colors, many paint stores have computers that can match a sample color. This capability is a real help if you are trying to duplicate a color that is outdated or one that has faded to a lighter shade over time.

Latex paints dry to the touch in about 30 minutes (compared to oil paints, which take 18 to 24 hours to dry to the touch), so you can apply several coats in rapid succession. They are water based, so cleanup is easy. Best of all, latex paints lend themselves to recreating the look that people love in an old finish.

Surface preparation

If you live in an old house, you should strive for a finish that complements the other materials—a gleaming modern finish would look out of place. You don't want the new work to draw the wrong kind of attention. Since most period woodwork was painted, the first thing is to understand to what kind of surface the paint was applied to.

The surface of most 18th-century woodwork prior to painting was as it was left by the plane. Sandpaper was not available, and other abrasives were reserved for the finishing of furniture. It was the practice of 18th-century craftsman to fair joints and remove minor tearout with burnishers and scrapers (see the photo below).

In the 18th century, a hand scraper often did the job of modern-day sandpaper.

Although painted, this 18th-century egg-and-dart molding shows subtle tool marks left by the carving gouge.

On a surface that is insufficiently sanded, paint will reveal machine marks and other evidence of 20th-century manufacture that can detract from the look of a traditional project.

If you look closely at period woodwork, you'll be able to see tool marks underneath the paint. At Drayton Hall, near Charleston, South Carolina, the original woodwork, dating to the 18th century, has miraculously survived with its single original coat of paint intact. Although the woodwork exhibits the finest craftsmanship, the carver's tool marks are evident through the paint (see the photo above). By 20th-century standards, 18th-century architectural surfaces are far from perfect. Tools marks were unavoidable and acceptable.

Am I recommending that you forego modern abrasives and power equipment to prepare your projects for paint? Not at all. Even if you are recreating a Colonial interior, I urge you to take advantage of the convenience and efficiency of 20th-century technology to prepare your wood surfaces. Just use them with restraint.

Modern woodworking machines leave telltale marks on your stock. They might escape detection before finishing, but they will be painfully obvious once the work is painted (see the bottom photo at left). That's why I recommend handplaning, scraping, or sanding to remove any trace of marks left by a table-saw blade, planer knife, or router bit.

Next, be sure the work is uniformly sanded, wiped with a damp cloth to raise the grain, and then lightly sanded again. Where you might normally sand to 220 or 320 grit, stop instead at 150 grit—a painted finish will require nothing finer. Give the woodwork a softer look by breaking the sharp edges with fine sandpaper. A light rounding will make your project look less than brand new; people will think it has been painted a few times already. Also, the slightly rounded edges will hold paint better.

Check your project all over for chips or dings and repair them. For most damage, you can use a sandable water-based wood filler. Damaged surfaces should be filled, leveled with a putty knife, and sanded again; excess filler can be removed with a damp rag.

Priming and caulking

New woodwork should be primed before finish painting. Primer is formulated to fill in and level a wood surface and eliminate minor irregularities that remain after sanding. Once it dries, primer can also be spot-sanded to further level and even out the wood. I recommend applying two thinned coats of primer ($\frac{1}{3}$ water to $\frac{2}{3}$ primer) instead of a single thick coat. This takes a little longer, but each coat dries faster and gives you more control over the material as you apply it.

Finally, use a premium-grade caulk to fill in any inside corners, small gaps, seams and joints in the woodwork (I use an acrylic adhesive caulk, such as Phenoseal, applied with a plunger-type caulking gun). Lay down a 24-in. to 36-in. bead long bead (any longer, and the bead might dry before it can be wiped), wipe it smooth with your finger, then clean off any excess with a damp rag. Caulking will give your project a softer look and eliminate any glaring hard lines. Besides filling in any small gaps, caulk will soften inside corners and make the piece look as if it's been around a while.

Basic painting

Latex paint can be brushed, rolled, or sprayed—let the type of work, the job site, and the size of the job dictate the best method of application. I get the best results when I thin out the paint ($\frac{1}{3}$ water to $\frac{2}{3}$ paint) and apply at least two coats.

When applying the paint, lay it down in long, smooth strokes, working with the grain when possible. Never go over an area with more than three passes, or the surface, which will dry quickly, might be marred by heavy brush marks. To get paint into all the quirks and crevices, you sometimes have to push it around into the corners and

across the grain, but the last pass should always be in the direction of the grain.

Antiquing techniques

Over time the interior woodwork of a house will take a beating. The effects of aging may be minor—simple discoloration, scratches, dents, worn edges—or severe. The edges might be completely gone, and the surfaces covered with nicks and dents. The paint might be heavily crackled and chipped, revealing other colors beneath. Contrary to what you might think if you are not a student of period woodwork, these conditions that indicate age and use are desirable to a certain degree. Today, an antique piece with multi-layered paint intact is more valuable than one stripped down to the wood.

Before you attempt any antiquing on your own new woodwork, examine a number of genuine antique surfaces (see the photos on p. 32). Get a sense of the normal wear patterns on woodwork and furniture, and don't get carried away. For instance, a corner cupboard won't have worn paint around the cornice; all the chips, dings, and scratches will be around the base. The paint on the cupboard might be sooty on the side next to the fireplace and faded on the side next to the window. Door pulls and handles

The way that paint ages depends on the type of paint, the thickness and number of applications, the compatibility of coatings, and ambient conditions, among other factors. That's why these three antique paneled doors look so different.

and the wood surrounding them might be bare from contact with human hands.

Over the years I've employed a number of techniques to "age" my projects using ordinary latex paint. Here are some you might want to try.

The layered look You can give your project an instant past by applying layers of different-colored paint—up to four different colors on a single piece. As the surface becomes discolored, worn, and chipped, the colors beneath will be revealed. Your project will look as if it had been repainted several times over the years. Woodwork with a paint history like this blends into an old house better.

Texturing, sponging, and stippling
In the 18th century, a painted surface was never smooth because early paint was thick and difficult to apply. In order to work it into the wood, painters used stiff brushes to push it around and into corners. The resulting surface was often streaked with raised brush marks. You can imitate the texture of old paint by going over a freshly painted surface with a semi-dry brush. When the paint dries, rub the surface here and there with steel wool. The paint will wear from the raised brush marks, leaving attractive striations.

For a varied surface texture, you can use a crumpled paper bag, scraps of sponge, hair combs, and sand. These materials, applied to the wet paint, create a texturally distressed surface. The color will remain intact, but it will have a more interesting feel. One favorite technique of mine is to stipple another shade, very close to the primary paint color, onto the paint surface. This mottled effect gives depth and

Not many people would want to recreate the heavily chipped and flaky look of an authentic old alligored surface (left). But a milder form (above) is more attractive. It can be created by applying lacquer over paint, followed by heat from a heat gun.

complexity to the paint surface and also imitates the look of old paint.

A crackle finish A crackle finish is a real challenge to reproduce. This is the crazed alligator finish of old paint. In the extreme it looks unattractive and cries out for a drink of paint, but in a mild form it looks old, fragile, and cared for (see the photos above).

Here's one method that I've found success with. It takes years to master in all its degrees, but it's worth the effort. After priming and caulking, apply two undercoats of latex paint. This will be the color that shows through between the cracks, so it should complement the top color. When the undercoat is thoroughly dry, coat selected areas with liquid gum arabic (a dried resin of the acacia tree, available at paint stores). Gum arabic is hygroscopic. If you allow it to dry for about an hour, then paint over it with your second color of latex, the gum arabic will absorb the water from the top coat, causing it to shrink and crack.

A crackle effect can be difficult to control, and it's nearly impossible to predict exactly what results you'll get. The thickness of the paint, drying time, the amount of gum arabic you apply, and the timing of the top coat— all these variables play a part. With practice, this technique gets easier, and the results get better.

Glazing As time passes, old furniture and woodwork accumulate dirt and wax in their corners and recesses. This dark buildup

This length of molding, possibly from an entry-door casing, is now on display at the Historic Charleston Foundation in South Carolina. Over the years, soot and dirt have accumulated, highlighting the subtly crackled paint and exquisite carving.

counterbalances the burnished areas, such as moldings and exposed surfaces (see the photo above). Glazing reproduces this effect and can age a piece 50 years in a few minutes.

Most people think glazing involves covering the entire piece with black goo, waiting a few minutes, then wiping most of it off. This approach produces an obvious effect without subtlety or interest. But what glazing is really about is creating highlights and contrast. You can use it to emphasize a surface texture, but like any other aspect of finishing, it must be used with restraint for the best results.

The basis of a glazing is the glazing mixture, sold at paint stores as glaze coat. This is a slow-drying vehicle with a slightly slippery feel. It stays wet and can be easily moved around a surface without penetrating into it and permanently altering the color. To this neutral-color mixture you add Japan color (a concentrated pigment). I generally stay away from black because it lacks complexity. I like to start with raw umber, which is dark enough to give you shadows that suggest layers slowly applied over time. If it's not dark enough for your taste, you can add more of it.

Apply the glaze with a soft brush, working it into corners and recesses. Wait about 30 minutes, then rub down the piece with a turpentine-dampened rag to remove the excess. What remains should look like the natural accumulation of dirt and wax.

Distressing Paint can be distressed by various methods of removal, ranging from lightly sanding the paint to burning it. Each of these techniques must be used carefully and logically.

The easiest technique is to remove paint with a variety of wire brushes, scrapers, steel wool, and sandpaper. Restrict the distressing to accessible edges, moldings, and outside corners; otherwise it will look phony. And use restraint. It is always easier to take a little more off than to repaint.

For burning effects you can use a small propane torch or a paint-stripping heat gun. Don't apply the flame directly to the wood at 90° and keep the heat source moving. Approaching the wood at an angle will give you more control and less risk of setting the piece on fire. After some experimentation you'll develop a repertoire of techniques that will allow you to create a range of interesting burn effects.

Heating and burning the paint create fragile, crusty blisters that flake away when scraped, leaving jagged craters. Using less heat raises smaller blisters to simulate proximity to a fireplace or other source of heat. Apply low heat for a few minutes over an area will discolor the paint in a convincing way and give it a tough, slightly bumpy quality.

ANTIQUING HARDWARE

If the wood looks old, so should the hardware. Here are a couple of simple ways to "age" inexpensive, off-the-shelf hardware.

ANTIQUING STAMPED-IRON HARDWARE

A good hand-forged hinge is mottled and textured in an attractive way. The outline is never straight, and its thickness varies. All in all, its appearance is irregular and slightly rough. A modern stamped hinge, however, which is supposed to resemble the hand-forged example, is perfect—perfectly straight and flat, and painted matte black.

Grasp the stamped hinge with a pair of tongs and heat it until it is red hot, either on a kitchen stove or with a propane and oxygen torch. Before it cools completely, lay it face down on a flat rock and strike it with a hammer to reshape its outline and texture the surface (see the photo below left), taking care not to strike the barrel of the hinge. When it becomes too cool to shape anymore, plunge it into a can of water to produce the variation in color. If the hinge needs more work, it can be reheated and reworked. The result (see the photo below right) is a hinge that is nearly indistinguishable from the real thing.

ANTIQUING BRASS HARDWARE

Some furniture and built-in cabinets should be fitted with formal brass hardware. Antique brass hardware has a smoky, somewhat tarnished look; brand-new hardware looks shiny and squeaky clean. You need to tone it down.

An easy method involves using ammonia to fume the brass. By fuming brass hardware you can achieve a range from a mellow gold to a dark brown color, depending on the exposure to the ammonia.

Begin by stripping the hardware of its protective lacquer coating. The easiest way to accomplish this is to soak the hardware in lacquer thinner. Let it soak all day and then remove any residue with fine steel wool. Be sure to remove all the lacquer, because it will retard the effects of the ammonia.

When the brass is clean of lacquer, pour about 1 in. of ammonia into a plastic bucket, and suspend the hardware above. Cover the bucket and place it in a corner of the shop, where it won't be disturbed. Check on the progress every three or four hours. The longer the hardware is exposed to the ammonia, the darker it will become. If it turns too dark, it can be polished with fine steel wool.

Hammering against a flat rock (right) imprints the texture and irregularity of the stone onto an H-hinge.

Both hinges at far right are modern, but reforging has given the one the left the look of an antique hand-forged hinge.

3 BOARD-AND-BATTEN DOOR

During the 17th and 18th centuries, the board-and-batten door was a common way to close one room off from another or to shut out the elements, and even today it's an appealing design. Whether you're working with a city apartment or an old farmhouse, a board-and-batten door is one of the easiest ways to introduce a strong period element. Its straightforward construction and sturdy hardware lend a room a casual early American feeling.

In a board-and-batten door, a number of narrow boards placed edge to edge form a single wide panel that spans the threshold. Two or three battens nailed across the boards hold them rigid and flat (see the drawing on p. 38). In period doors, the edges of the boards were either planed with a tongue and groove or shiplapped, then beaded, chamfered, or otherwise decorated. The joint kept the boards in alignment and accommodated the seasonal shrinkage and expansion across their width. The decorative treatment concealed this wood movement and added visual interest to an otherwise plain slab of wood.

DESIGNING THE DOOR

A board-and-batten door introduces a strong vertical element, providing relief and interest to a room oriented on a grid. To add to this interest, you can alter and embellish the basic design. When I build these doors, I usually vary the widths of the individual boards. If I need a 30-in.-wide door, I might design it with 11-in. planks at the edges and an 8-in. center plank. This brings the two lines of beaded decoration closer together, reinforcing the verticality of the door and creating an interesting rhythm.

Most early doors have two battens, but some have three or even four, with the edges decorated in a variety of ways (see the drawing on p. 39). Some were cut square, but this produces an effect at odds with the door and somehow not an integrated part of it. A simple embellishment was a long chamfer along the edges and the ends, cut with a bench plane or even a wide chisel. This was a common treatment since it required no special tools. Often the edges were molded with a beading plane. A ⅜-in. bead is a simple detail that softens the batten and creates a pleasing shadow. The most elaborate treatment was to mold the edges of the batten with a complex plane. This was a rare and fine detail that is somewhat luxurious compared to the simple construction of the door itself.

MATERIALS

In the 17th and 18th centuries, pine was the wood most often used for interior woodwork. It was plentiful, available in wide planks, and easy to work. Today, however, the cost of the best grades—clear and #1 pine—is high and rising every day. If you can't afford pine, poplar is an acceptable substitute in terms of color, grain, workability, and price. Or you may be able to use a lesser grade of pine, if you pick your boards carefully. At my local lumberyard, I frequently go through the bins and cull the better #2 boards, some of which are nearly clear except for some very small knots.

When the door is completed, it must hang well, meet the door stop, and operate easily, so try to select boards that are straight and flat, without any wind or cup. Avoid boards with large knots, pitch pockets, or other defects that will be difficult to cover with paint.

Whenever possible I design doors to be built using dimension lumber. Here is where the woodworker gets a break. Working with ¾-in. stock will save you time and effort—you

A TYPICAL BOARD-AND-BATTEN DOOR

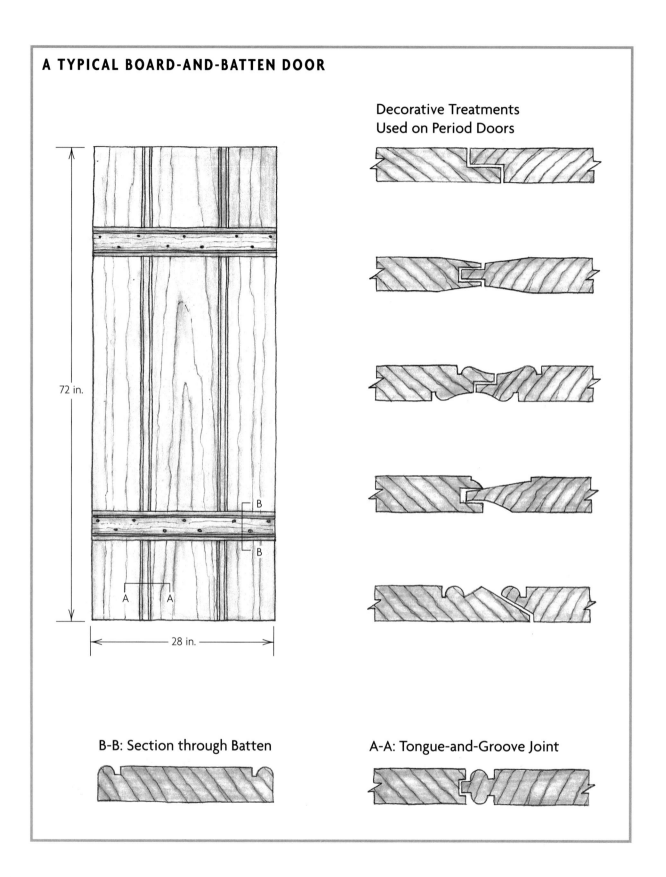

Decorative Treatments
Used on Period Doors

72 in.

28 in.

B

B

A   A

B-B: Section through Batten

A-A: Tongue-and-Groove Joint

EDGE TREATMENTS FOR BATTENS

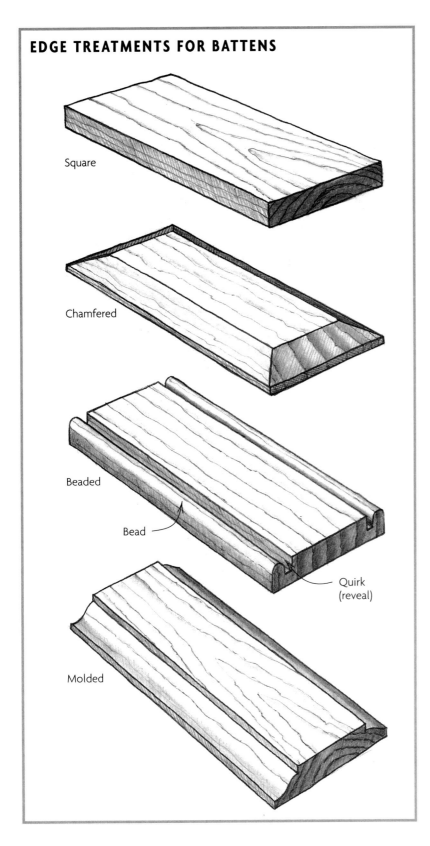

Square

Chamfered

Beaded

Bead

Quirk
(reveal)

Molded

don't have to thickness-plane it, since period interior doors are often less than 1 in. thick. I've made many board-and-batten doors with ¾-in. pine, which looked and worked just fine.

MILLING THE BOARDS

Joint a straight edge on each board and then rip them to width, leaving an allowance for the tongue portion of the joint. Cut the boards a little longer than their finished length so that any snipes or blips at the ends caused by an errant router can be trimmed off later.

The boards making up the door in this project are joined with a tongue-and-groove joint. This interlocking joint functions in two ways: It registers the surfaces of the two boards on the same plane and it also accommodates the inevitable shrinkage and expansion of the wood.

During the 18th and 19th centuries, the tongue-and-groove joint was cut with tongue-and-groove planes. These handy planes, found in nearly every carpenter's toolbox, came in various sizes. You can still get them at auctions or from tool dealers, but they're becoming scarce.

Today, of course, matched router bits have replaced matched tongue-and-groove planes. They

A marking gauge is best for laying out machine cuts. It makes a clear, visible line, so fewer test cuts are needed to set up a cut.

Matched tongue-and-groove router bits produce a snug joint that fits perfectly.

function the same way as planes and offer nearly the same results with the advantage that they can cut material almost regardless of grain direction. The main difference between the two is that the planes cut in a linear direction, whereas routers make a rotary cut. The plane slowly forms the profile by repeated passes against the wood's edge, so any deviation is usually spread out over a 6-in. to 8-in. length of molding and is difficult to detect. In contrast, a router error appears as a blip, a short abrupt arc, that corresponds to the diameter of the router bit. Router errors are often highlighted by an accompanying burn mark. They are easy to spot and are difficult to remove or hide.

Routing the tongue-and-groove joint

Begin by laying out the joint. For this I like to use a marking gauge (see the photo above left) because the clear scribe lines make it easier to set up the cuts.

It doesn't matter which side you cut first; tongue or groove. Both bits of the set are designed to cut perfectly mating halves, and each bit has a ball-bearing guide to guarantee a uniform depth of cut (see the photo above right). Only the position of the bit in relation to the thickness of the material must be set.

On this project, it would be difficult to run the large boards over a router table. It is easier and also safer to use a hand-held

router. I recommend performing test cuts on some scrap before milling your door stock. Securely clamp your material to your bench to make sure the edge of the board clears your workbench, allowing the router bit to cut to its full depth; otherwise the ball-bearing guide might run along the edge of the workbench instead of the workpiece.

A hand-held router's bit spins clockwise as it cuts, so you should move the router from left to right for a smooth and safe router operation.

Making the bead

The bead is probably the most common embellishment found on early woodwork and the easiest to perform. It can be cut with either a beading plane, a router bit, or a scratch stock. On a door that's ¾ in. thick, a ¼-in.-dia. bead would look good. Alongside the bead profile, the router bit will cut a small reveal called a quirk. On early moldings, the quirk comes to a point; later beads have flat-bottomed quirks. If you prefer the earlier effect, I suggest you cut the molding with a scratch bead because almost all router bits and most molding planes found today will produce a flat-bottomed quirk. The bead could be cut on the groove portion of the joint, but that would weaken the joint. It's better to cut it on the tongue portion of the joint.

MILLING THE BATTENS

The battens are made up of ¾-in. pine about 4 in. wide. The fastest and easiest way to cut the bead and quirk on the edges is to set a ⅜-in. beading bit into the router table and align the fence with the ball-bearing guide. Set the height of the bit so that a full bead, without any flat spots, is cut on the edge of the batten. Run the pieces over the router table smoothly, without a pause. A deliberate and steady pass will result in a clean, full bead without any blips or burn marks to clean up later.

ASSEMBLING THE DOOR

When all the milling is done, assemble the boards to make sure that the tongue-and-groove joints fit properly (see the photo below). If you encounter a few tight spots along the length of the joint, don't force the joint together. Instead, use a shoulder plane on the tongue portion of the joint. If the door must be moved around the shop, clamp temporary battens across the ends of the planks to hold them together.

When all the boards fit according to plan, you can lay out the location of the battens. I usually place them about 12 in. from each end (see the photo on p. 42). (see the photo on p. 42) In period doors, the battens were laid at 90° to the vertical boards and secured with wrought-iron rosehead nails driven through the batten and the vertical board, then clinched over on the door's face side (the side without the battens). Today, you can purchase hand-wrought iron nails made by a blacksmith, but they are expensive. Tremont Nail Company sells a "rosehead" soft nail just for this application for $6.40 a box that contains about 50 nails. For a board-and-batten door, you'll probably need

The tongue-and-groove joints for the board-and-batten door should fit snugly, but not require excessive force to assemble. When the tongue is fully engaged, there should be a reveal of ⅛ in.

The battens are normally placed about 12 in. from the ends of the door and are temporarily held in place with handscrew clamps. Here a framing square is being used to position the batten at 90° to the edge of the door.

nails that are 2½ in. long. Each vertical board receives about three nails placed in a triangular pattern for strength.

Predrill the holes for the nails, then use a small sledge or a 2-lb. mason's lump hammer to drive the large nails through the batten and the door. The 18th-century carpenter would drive the shank of the nail through a small leather washer (see the top photo on the facing page), which prevented the nail from backing out as the pointed end was clinched over on the other side. Later, any metal protruding above the surface could be knocked flat with the lump hammer or countersunk with a

nail set, then filed flush. The nail heads were left proud. Sometimes the excess leather was cut away, but I've seen many 18th-century doors with the original leather washer peeking out from beneath the nailhead.

If you want to forego period authenticity, you can also fasten the battens to the boards with screws. The advantage of using screws is that they can be countersunk (the holes can be plugged or filled with a mixture of sawdust and epoxy), and the face of the door will remain unblemished. This might be a good method if you plan to leave the door unpainted.

HINGES

On the doors I make I often use wrought-iron strap hinges that about 14 in. long, but H hinges, which cost less, are also historically correct and very attractive (see p. 24). High-quality period reproduction hinges and other types of hardware are available from various well-known catalog suppliers and a growing number of small foundries that turn out interesting and unusual items, as well as custom orders.

H hinges don't require any mortising so they are easily mounted onto the surface of the door with either rosehead nails or the screws that come with the hinges. These screws usually have a fake nailhead pattern over the Phillips slot. When in place, their irregular faceted head looks almost like the real thing. For the finishing touch, you can cover the slot with cold solder or auto-body filler, then paint it to match. When completed, it will look just like a period nail head.

For better holding power, 18th-century carpenters drove their nails through small leather washers. The washer held the nail fast and prevented it from backing out of its hole when it was clinched.

HANGING THE DOOR

Plane and trim the door to fit the opening, and attach the H hinges. Remember, this is effectively a solid slab of pine, and it will expand and shrink seasonally, so you need to allow about ⅛ in. per 1 ft. of width to accommodate the movement. To cover the shrinkage and keep the door airtight in the winter, use a ⅝-in. doorstop around the jamb.

When the door fits, place it in the doorway. Using tapered shims (shingles work well) to hold it in place, check for even spacing all around (see the photo at right). Then drive two nails (or screws)—one for each hinge—into the casing. With the door secured, remove the shims and test the door. If it swings without binding and there are no unsightly gaps, drive in the rest of the fasteners.

Tapered cedar shingles are used to shim the door into position in the opening. A gap of at least ⅛ in. all around will ensure easy year-round operation of the door.

TONGUE-AND-GROOVE PLANES

During the 18th and 19th centuries, when narrow boards were joined to create larger surfaces, their edges were often prepared with a tongue-and-groove joint. This sturdy joint was meant to be concealed, but when done well, it is attractive and becomes a design feature (particularly when seen from the end grain). Tongue-and-groove joints were used on floor boards, tabletops, drawer bottoms, cabinet sides, and, of course, board-and-batten doors.

The joint was often cut with a pair of planes: one to cut the tongue and the other to cut the groove (see the photos on the facing page). These planes were always sold and maintained as a matched pair; the random pairing of two similar but unmatched planes would produce a joint that might not fit together easily and a surface that might not lie flush. Tongue-and-groove sets were sold to cut a joint centered on edge, and were designated by the thickness of the wood being cut. The range was from $\frac{1}{4}$ in. to $1\frac{1}{4}$ in., with the most common size being $\frac{3}{4}$ in.

The most common style of plane was the unhandled type, measuring $3\frac{1}{2}$ in. high by $9\frac{1}{2}$ in. long. For heavier work on thicker stock, some planes came handled for comfort and greater control. For the carpenter on the move, there was a combination plane, which cut a tongue when pushed in one direction and a groove when pushed in the other.

Tongue-and-groove planes have a built-in fence and depth stop, so aside from the blade, there's nothing to set up or adjust. To set the blade, hold the plane upside-down and sight down the sole from the toe end. The profile of the blade should be in perfect alignment with the profile of the sole, and the cutting portions of the blade's profile should project uniformly to ensure an even cut and good chip ejection.

These planes are fairly easy to use. If you keep the plane upright and pressed against the outside edge of the board as you push the plane forward, it will cut until the depth stop rides against the surface of the board and prevents the plane from descending deeper into the cut and removing more wood. Tongue-and-groove planes are one of the few types of plane that are held at 90° to the wood's surface. (Most molding planes are "sprung," or held at a slight angle to the wood, as described more fully in the sidebar on pp. 55-57.

A matched pair of tongue and groove planes produces a perfect fit between these two boards (left). The tongue-cutting plane (below left) has an integral fence that positions the tongue at the center of the board. The center portion of the blade is recessed into the body of the plane, stopping the plane from cutting below a specific depth. The narrow groove-cutting blade (below right) is centered and supported on a metal skate attached to the body of the plane, ensuring perfect placement of the groove in relation to the tongue.

FOUR-PIECE BASEBOARD

When I first entered the front room of my old farmhouse, I was amazed by the size of it. By period standards, it was a big room—22 ft. square. All the doors and windows were framed by elaborate built-up moldings with carved pinwheel rosettes at the corners. The doors were paneled, and each panel was surrounded by delicate moldings. It was evident that a lot of work and care had gone into this part of the house. This grand display was carried on throughout the room, right down to the elegant baseboard. When I examined it more closely, I could see it was simply a wide pine board with a molded rabbet, capped with a ⅝-in. bullnose and a slender cove. It was simple, but bold and handsome. For the effort involved to make it, there was a big payoff.

The distinguishing characteristic of period baseboard is its height. Naturally, there should be a relationship between the height of the baseboard and the rest of the room. A grand Georgian mansion with 12-ft.-high ceilings might have baseboards 12 in. tall. In a simple Cape Cod cottage with 7-ft. ceilings, the baseboard might be only 6 in. tall. I have rarely encountered baseboard molding less than 7 in. tall in 18th- and 19th-century houses, unless it was a later replacement.

Period baseboards can range from a simple square-cut or chamfered board to a stepped-down or rabbeted main board, capped with a generous crown. Today, a woodworker can pop a bit into a router and without regard to grain direction or the hardness of the wood at hand, zip from one end of an 8-ft. board to the other in about two minutes. The result will be a crisp and uniform length of molding. In the 18th century, however, producing a molding was a slow and laborious task (see the sidebar on pp. 55-57).

Then, as now, baseboard was installed where the walls meet the floor, and its primary purpose was to protect the wall at this vulnerable junction. But in the 18th century, baseboard (and other trim as well) had another function as well—it served as a guide for plastering. On an old house, the trim appears to be almost buried in the plaster. This softened look is entirely in keeping with the technology of the 18th century (see the sidebar on p. 48) and one you should strive to maintain.

Even in a modern room, a tall period baseboard can lend a classic touch. You might imagine no one notices baseboards, but that's not true. If you were to compare two otherwise identical rooms in a modern house, one with 3-in. clamshell everywhere and the other with taller, more elaborate baseboard, you'd instantly see the difference. Period baseboard slows down the eye, making the visual trip around the room more leisurely. It also makes other details in the room more noticeable and important.

DESIGN

In baseboard moldings, the guidelines regarding chronological styles aren't very rigorous, so you can experiment with different profiles and see what looks best to you. As you consider different styles, keep in mind that period moldings were illuminated by dim candlelight. To be seen at all, moldings running along the floor had to be bold and cast strong shadows. Generally, as with other types of woodwork, the fancier, more complex baseboard treatments were reserved for the front downstairs rooms. On the second floor, the woodwork tended to be simpler.

If your design calls for a very tall one-piece baseboard molding, it will have to be made up by a custom millwork shop because suppliers today don't stock moldings more than 6 in. or 7 in. tall, and

HOW OLD WALLS WERE PLASTERED

Modern trim is installed over drywall, so its lines are sharp and crisp. Traditional trim is installed directly over lath (see the photo at right). In the 18th century, all the trim (window and door casings, chair rails, mantels, and baseboard) was installed before plastering. Unlike modern timber-framed structures, 18th-century house frames did have studs supporting the wall (see the drawing on p. 7), and since these studs were often roughly cut and shaped from air-dried wood, they never lined up perfectly straight. Their edges had hollows and high spots. The split lath nailed to the studs also presented an uneven surface. All of this caused problems for the plasterers, who had to apply plaster coats of varying thickness onto the lathe in order to achieve a flat and smooth wall. To accomplish this

quickly and easily, they used the moldings as a guide to gauge the required thickness of the plaster at a particular area of the wall. This thickness could vary considerably —as much as 1½ in. from one side of a window to the other, as

I discovered when I restored the plaster in my front room!

On the finished wall, the moldings projected a uniform amount from the plaster wall, disguising the actual thickness of the rough and finish plaster.

In centuries past, baseboard was nailed directly to the studs or lath before the wall was plastered. It served as a guide for the plasterer; that's why old baseboards are partly buried in the walls.

with good reason. Wide baseboard moldings require more wood to produce, so they cost more. And because of their thickness and width, they are liable to twist and warp in storage, rendering sections of the stock

useless. It's much easier and cheaper for a lumberyard to stock smaller, narrower profiles.

So how does someone reproduce a wide molding without spending a small fortune? One way is

to use standard #2 pine, cut square and available in lengths up to 16 ft. for the body of the baseboard, then trim it out with smaller stock or shop-made profiles. That's how I reproduced the baseboard molding for this

The baseboard pattern in the 1812 section of the author's house. Stepping the sample shows how a complex antique pattern can be recreated with modern stock moldings.

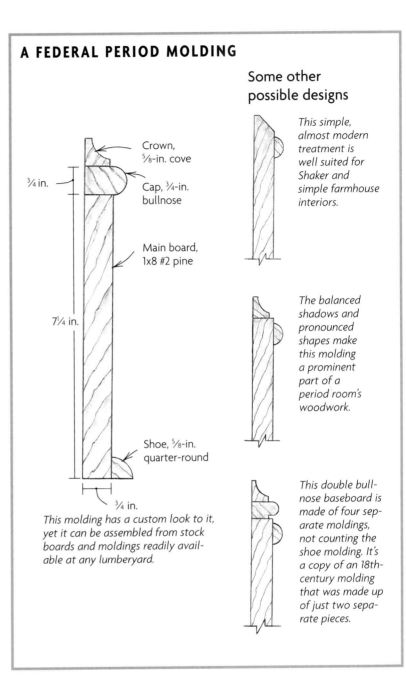

A FEDERAL PERIOD MOLDING

Crown, ⅝-in. cove

Cap, ¾-in. bullnose

¾ in.

Main board, 1x8 #2 pine

7¼ in.

Shoe, ⅝-in. quarter-round

¾ in.

This molding has a custom look to it, yet it can be assembled from stock boards and moldings readily available at any lumberyard.

Some other possible designs

This simple, almost modern treatment is well suited for Shaker and simple farmhouse interiors.

The balanced shadows and pronounced shapes make this molding a prominent part of a period room's woodwork.

This double bullnose baseboard is made of four separate moldings, not counting the shoe molding. It's a copy of an 18th-century molding that was made up of just two separate pieces.

project (see the photo above), which was copied from my own home. It was the original molding and was used generously on both the first and second floors of the 1812 section of the house (see pp. 14-17).

INSTALLATION

The baseboard you are about to assemble is composed of four parts: the main board, the cap, the crown, and the shoe (see the drawing above right). The main board is the largest section, and its dimensions dictate the overall effect of the design. The cap is the boldest segment; it creates a strong shadow line and gives the baseboard definition. The crown creates a softer shadow and provides a transition from the cap to the wall. The shoe, a small quarter-round molding that visually anchors the baseboard, adds interest to the profile and protects it from damage. It also hides small gaps between the baseboard and the floor.

For the main board use #2 Select New England white pine. At the lumberyard, carefully pick through the bins yourself, choosing the cleanest and straightest pieces of 1x8 stock you can find. Knots should be limited to ¾ in. in diameter, with no more than two per linear foot. The grain patterns should be mild and regular across the width of the board. Reject boards with erratic or wild grain.

For the cap, use a sleek ¾-in.-thick bullnose molding with an overhang of ⅜ in., so the bullnose will fully clear the board. Stock windowstop molding can be used, or you can make your own cap in the shop with a router or shaper.

For the crown, a ⅝-in. cove will be easy to install and give you a nice transition. This small molding will be substantial enough to hide large gaps, and its slender profile will be flexible enough to work nicely along the wall without much effort.

This baseboard doesn't really need a shoe molding, but it adds some visual heft and performs the practical function of covering gaps between the board and the floor (see the photo above right). In an old house, shoe molding also cuts down on cold-air infiltration from foundation walls and cellars.

Shoe molding protects the baseboard and hides small gaps between the baseboard and the floor. It also helps stops drafts from foundation walls and cellars.

Tools and methods of work

To put together a baseboard, you don't need much in the way of tools. You can get by with just three: a chopsaw, a block plane, and a jeweler's saw.

A good chopsaw that can crosscut the edge of a board at 45° is essential. Fitted with a 60-tooth crosscut trim blade, it will produce perfect silky cuts without tearout. Get a saw that uses a 10-in. blade; these blades are readily available and also go on sale frequently. The other blades (8½-in. and 12-in. dia.) might not be regularly stocked items.

A block plane is handy for trimming debris from miters or for making small adjustments in the fit of the moldings. A low-angle block plane produces the cleanest finish on end grain. I like the Lie-Nielson #102 and the #60½ block planes. Although they're fairly expensive tools, they cut cleanly and produce excellent results. Keep in mind that the more elaborate your trim work is, the closer inspection it will receive. So when seen up close, the joints had better be good and snug. If your budget is tight, there are many less expensive planes that will perform adequately.

A jeweler's saw is great for coping small crown moldings. Lighter in weight and using finer blades than a regular coping saw, this fine-cutting saw will leave clean edges that rarely need sanding.

When it comes to installation, you have two choices. If you're a die-hard traditionalist, installing your trim by hand using a hammer and nails is the time-tested way of doing the job. For attaching the main board, 6d finishing nails will probably suffice, but keep a few 8d nails for any stubborn, hard-to-attach areas. For general work, a 12-oz. hammer is fine. It's heavy enough to drive finish nails, but light enough not to bang up your work. Once the nail head is hammered to within $\frac{1}{16}$ in. of the wood surface, you'll need a nail set to drive the head neatly beneath the surface.

If you're a small contractor or a serious hobbyist, or if you are renovating an entire house, you should consider acquiring an air compressor and a pneumatic nail gun (see p. 27). An air-powered finish nailer (see the photo above right) leaves an almost invisible entry hole and virtually eliminates the splitting of small moldings. You can easily cut your installation time in half using pneumatic nailers and probably improve the quality of your work as well.

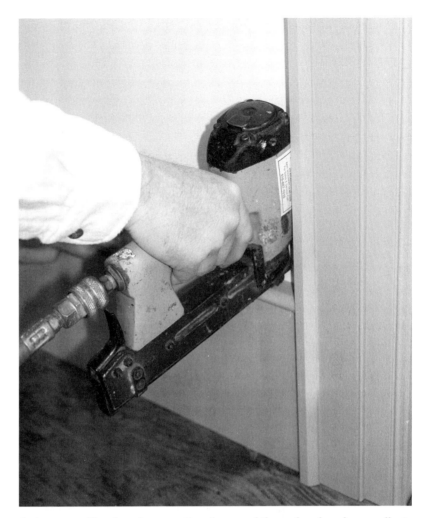

On long runs of baseboard, pneumatic nailers speed the work and produce excellent results. The gun shown here is powered by a portable $\frac{3}{4}$-hp compressor and shoots nails ranging in length from $\frac{1}{2}$ in. to $1\frac{5}{8}$ in.

The main board

The easiest part of the baseboard to install is the main board. If the baseboard is being installed before the floor, just cut the baseboard sections to length and drop them in place. The inside corners can be butted instead of mitered. The outside corners should be mitered at 45°, then glued and nailed.

Splicing will be unavoidable, since some walls will be longer than your baseboard pieces. The trick is to execute a splice that doesn't attract attention. For the best appearance, the seam of any end-grain joint should be fair and almost invisible. To achieve a joint that is easy to nail and camouflage, cut each end at 45° and overlap them. After nailing,

A properly designed baseboard should die cleanly into the the door casing.

sand and level any high spots, then caulk the joint. Staggering the parts of the baseboard will also make the joint less conspicuous.

At door casings, generally the baseboard dies into the back of the backband molding that is attached to the casing's outside edge (see the photo above). This treatment is easy and produces a neat appearance that doesn't cut the baseboard off at an odd angle or allow it to extend past the door molding.

If the floor is already in place it might be necessary to scribe the baseboard to the floor. Since any small gaps will be covered by the shoe molding, only those greater than ¼ in. will need attention.

To scribe, level the section of baseboard being installed. This should reveal the extent and location of any gaps. Next, set the legs of a small compass to the widest gap, and drag the compass with the pointed end along the floor and the pencil along the baseboard. The line on the baseboard should mirror the surface of the floor. Using a jigsaw, spokeshave, or block plane, trim the baseboard to the line.

If the main board lies flat against the wall, secure it to the wall with 6d finish nails. Outside miters should be lightly glued and nailed. Don t worry about small gaps—these can easily be filled later.

If the wall has some wind that prevents the main board from lying flat against it, countersink some #8 screws 1½ in. long through the board and into the studs or lath behind it; if they don't catch hold, try 2½-in.-long screws. If possible, place any screws where they will be covered over or partially hidden by crown or shoe moldings.

The cap

This large bullnose molding is a simple half-round shape that is very forgiving. Since it presents no hard lines or shadows, you can fudge the overhang if needed or even reshape the profile at the corners for a neat miter.

The cap molding can be mitered or coped. Since only the inside miters can be coped, some woodworkers choose to miter everything. The only advantage to coping small moldings is that some profiles match up better than others when coped, and any discrepancy in a corner angle is not as noticeable. I sometimes find that caulking is easier and looks better on a coped miter. But since a cove will be placed over the bullnose, it doesn't really matter which method you choose.

To cope a section of square-cut molding, make a 45° cut to reveal a clear outline of the molding profile. Then, with the workpiece secured in a vise, cut to the outline with a jeweler's saw, undercutting slightly. Minor tearout and mistakes can be filed or sanded out. The profile of the coped piece will interlock with a section of the same molding that meets it at 90°. The photos on the facing page show the sequence of events.

Coping Small Moldings for a Seamless Fit

A jeweler's saw cuts the outline of the profile, revealed by a 45° cut on the end of the piece (left). The profile of the coped piece will interlock with a section of the same molding that meets it at 90° (below left and right).

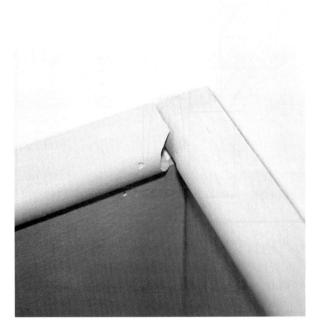

The crown

The crown is a small transitional molding that sits on the cap molding and against the wall. This important component softens the joint and provides additional visual interest. It also helps to hide any small gaps between the cap molding and the wall.

The shoe

Shoe molding is rarely used today, but I use it whenever I can in my work because it makes a baseboard look substantial. it also performs the practical function of protecting the baseboard and covering any gaps between the floor and the main board.

FINISHING TOUCHES

Caulking and filling the baseboard are critical steps in preparation for painting (see p. 31). If you're aiming for an 18th-century look, every dent, nail hole, seam, and joint must be filled in. The baseboard's appearance will be softened (see the top photo at right), and the subtle shadows cast by the individual moldings will blend. This important step unifies all the baseboard's components and gives the impression of one-piece antique woodwork that has been painted many times over the years (see the bottom photo at right).

A good caulking job eliminates hard seams and gaps.

The completed baseboard casts a strong shadow along the floor, adding interest and detail to the room.

MOLDING PLANES

The production of molding prior to the invention of spindle shapers and molders was an exercise that called for patience, hard work, and well-honed plane irons. During the 18th and 19th centuries, furniture makers and carpenters used molding planes. Like router bits, these simple tools are limited to a single profile. So, for every fancy molding, a carpenter needed a separate plane.

Most molding planes are oblong tools, about 3½ in. tall by 9½ in. long (see the photo below). Before 1830, molding planes were made of woods such as apple, cherry, and yellow birch (a wood that resembles a dense mahogany, not the birch we know today). After 1830, molding planes were made of beech, a stable wood that's easy to work, with a density similar to that of oak. I've made close to 100 planes of beech, so I can understand why it was considered the ideal wood for planes.

Molding planes have three critical parts: the body, the iron, and the wedge (see the drawing on p. 56). The body is a solid piece of wood with a notch (the throat) cut diagonally through it. The angle this notch makes with the bottom (the sole) of the plane determines the cutting angle of the iron. This varied according to the kind of wood that was to be cut. For hardwoods, the iron was bedded at 55°, called the middle pitch. For dense figured wood, 60°, or the half pitch, was used. The most common bed angle, 50° (the York), was used for architectural work and softwood. The wedge is tapered to fit into the throat. It had two functions: to hold the iron firmly against the bed (the back of the throat) and to facilitate the ejection of the shavings. Most molding planes have an integral non-adjustable stop and fence to ensure the production of uniform molding. The plane is simply held to the edge of the

(continued on p. 57)

The molding plane is a simple tool that can produce a complex profile.

The parts of a molding plane

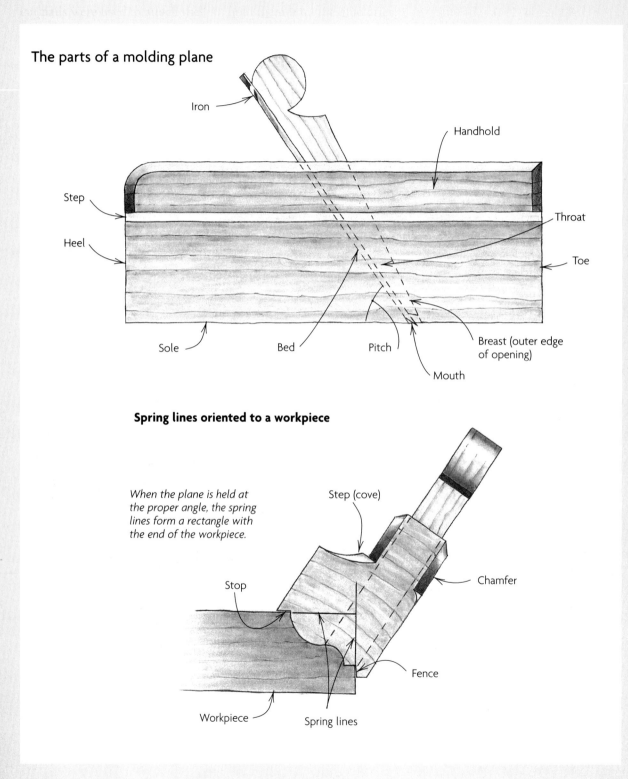

Iron

Handhold

Step

Throat

Heel

Toe

Sole

Bed

Pitch

Mouth

Breast (outer edge of opening)

Spring lines oriented to a workpiece

When the plane is held at the proper angle, the spring lines form a rectangle with the end of the workpiece.

Step (cove)

Chamfer

Stop

Fence

Workpiece

Spring lines

(continued from p. 55)
board by the fence and pushed until the full profile is formed, at which point the plane stops cutting. This is a simple design, and it works beautifully.

USING A MOLDING PLANE

A molding plane cuts in a straight line. As you guide it carefully along the edge of the workpiece, it removes the wood in the path of the lowest projecting portion of the blade. With every pass, a little more wood is taken off. Eventually, the plane descends into the wood, and the full molding profile is rendered. Cutting a simple half-bead might require six to eight passes with a molding plane. A more complex profile might need twice that many passes. Each pass has to follow exactly in the the tracks of the previous passes; any shifting of the plane would alter the intended molding profile.

On the front end of a molding plane (the toe), you often find a pair of perpendicular incised lines, called spring lines. The longer vertical line is visually aligned with the edge of the board (see the photo at right). This places the plane's body at an angle to the board. The reason for this is that as the profile is formed, the plane has a tendency to roll off the board. "Springing," or angling, the plane into the edge of the board makes it easier to keep the plane on track, and the resulting molding will be properly formed and uniform.

It is commonly believed that early woodworkers set themselves at one end of the board, took a pass with the plane, then returned to the opposite end to begin another pass. That just doesn't make sense to me, since this practice would have resulted in a lot of extra walking with nothing to show for it.

When I plane molding, I start about 30 in. from the front end of the board and take enough passes to nearly complete the profile. Then I step back another 30 in. and repeat the process until I reach the end of the board. Only when I reach the rear end of the board do I take a full-length pass with the plane in order to integrate the sections and fair any blips or differences in the profile.

Today, using molding planes is regarded as an arcane woodworking practice. Some people feel that these tools have no place in the modern shop, and from the standpoint of speed and efficiency, they are probably right. But using a molding plane is a satisfying experience, one that allows you to encounter the unique pleasures of hand-tool woodworking and puts you solidly in touch with the history of our craft.

The molding plane is a simple tool that can produce a complex profile.

5
WIDE-PLANK FLOORING

One of the first things I look at in an old house is the floors. Beautiful dark, wide boards studded with hand-wrought nails and worn smooth by countless footsteps to a mellow patina grab my attention every time. Nothing conveys a sense of history in a room more effectively than an old floor. In a newer home, a plank floor can take a room right out of the cookie-cutter category and add a sense of warmth and intimacy that many modern houses lack.

During the 17th and 18th centuries, flooring was generally laid double thickness on the first floor of the house and single thickness on the second floor and the attic. The reason for this had to do with the square-cut floor boards that were used. Since the edges of floor boards didn't interlock in any way, cold and dampness could easily pass from the cellar to the first floor. To form a barrier to the cold, a subfloor of roughsawn boards about ½ in. thick with irregular edges (this was called slit-stuff) was laid down at 90° to the floor joists. On top of the subfloor, the finish floor was laid, running in the same direction as the subfloor and fastened with hand-forged nails.

Since the flooring was inconsistent in dimension and the supporting joists so randomly arranged, the floor patterns in old houses can be quite surprising, at least to the modern eye. Sometimes the length of the boards will almost span the room, with one or two planks running along the far end at 90° to the rest of the floor. I've also seen floors laid with a single end seam running down the middle of the room (see the photo below), where no effort was made to stagger the seams. Maybe there was once a partition located over the seam, or maybe laying boards this way was just easier than staggering the ends.

SELECTING THE MATERIAL

The earliest floors were made of oak, following the English tradition. During the 18th century, pine was commonly used for flooring, and for good reason. Pine trees were large and plentiful, and the wood was relatively easy to cut, joint, and surface. These same qualities make pine the most suitable wood for reproduction floors today.

Raised knots and a worn surface lend charm to this old pine floor, whose wide boards were laid with their butt ends in a line, creating a single seam across the entire room.

Shiplapped pine floor boards are commonly available in thicknesses of 7/8 in. and in widths up to 12 in.

A number of companies offer recycled antique boards of southern yellow pine or eastern white pine salvaged from old buildings and barns. These choice boards are available in widths up to 22 in., and they make for a floor of incomparable color and richness. The only drawback to these beautiful boards is their cost.

A more economical alternative is locally milled lumber. There are many mills that produce boards up to 12 in. in width at a fraction of the cost of recycled antique boards. This material (see the photo above) is sold as #2 pine shelving with shiplapped edges and square ends. I've had good luck buying floor boards locally—the boards have been uniform in quality, dimension, moisture content, and color.

People often think that a pine floor will be soft. Well, that's true, it is soft. My own kitchen floor is lightly pocked with my wife's heel marks, but overall the effect has improved the floor. In the years since I laid that floor its color has deepened to a warm pumpkin-like shade, and regular polishing has brought up a beautiful soft luster. The tiny dents are the perfect finishing touch. That floor now looks as if it's been down for 200 years.

The advantages of a prefinished floor

A wood floor must be sealed and protected from dirt, spills, and traffic. This can be done before or after the flooring is installed. Generally, flooring is laid down unfinished, then sanded with a heavy drum sander, which raises

clouds of dust. After the rough sanding, cracks and open seams are filled with putty. Then the floor is finish sanded, and a stain is applied, followed by a protective floor finish, such as polyurethane. Most contractors who sand unfinished flooring will insist on countersinking any nail heads at least 1/4 in. below the surface. So much for your nail-studded floor!

An alternative approach is to prefinish the top surface of the floor boards before laying them down. I've used this method successfully, and it has several advantages. You eliminate the dust, you don't have to pay a contractor to do the sanding, and you get a great-looking floor. It's the method I'm going to describe here.

Obtaining and preparing the floor boards

Wide-board flooring is generally sold by the board foot. Once you find the square footage of the room, add another 10% to 15% to account for the wood that will be lost to split or warped boards, rough ends, and other defects.

If you order flooring from a lumberyard, you'll most likely receive a mix of boards ranging in lengths from 8 ft. 16 ft. Once you get the lumber off the truck and into your house, group the boards into piles by length: 4 ft. to 8 ft., 8 ft. to 12 ft., and 12 ft. to 16 ft. The boards you get will have been kiln-dried, but then stored in an open shed or warehouse without temperature control. They will need to acclimate to the conditions in the room where they will be installed. For that reason they should be stacked and stickered (separated by thin strips of dry wood placed between them) so air can circulate around them.

Let the floor boards season for at least six weeks before you install them. This is a critical step in the preparation of your flooring material. Well-seasoned boards that have been tightly laid will open about ⅛ in. at the seams. Small gaps are acceptable and should be expected. Flooring that has been improperly stored and prematurely laid down may shrink as much as ½ in. between the boards. These large, unsightly gaps can ruin an otherwise beautiful floor.

Before milling, cull the longest and best-looking boards from each of the stacks to use where they will best be seen. By the same token, put aside any short or knotty boards to lay under built-in cabinets and inside closets. It's also a good idea to eliminate obviously defective material by cutting off any ugly areas and end sections with large splits. Also remove loose knots. Boards with a little twist, wind, or crook should be cut down into shorter lengths to minimize the problem.

MILLING

Floor boards come rough on one side and relatively clean and smooth on the other, but the clean side often has planer marks, grade stamps, and small rough patches. That's because the flooring manufacturer is assuming the surface will be sanded after the floor is installed. If you prefinish your floor, you'll need to remove these blemishes first by running the boards through a planer. A small 12-in. planer (see the photo below) is what I use—it's portable and relatively quiet, and it leaves the boards ready to finish. Since the object of planing is just to clean and smooth the surface, you need only take a light cut off

Two or three passes through a small planer prepare the floor boards for finishing.

the top. Do this in two or three passes—one or two light cuts and a final very fine cut.

Because some pine floor boards have an irregular grain and the direction of the grain changes around knots, a clean, smooth top surface may be difficult to achieve. The best way to get a smooth surface is to make three passes. First make a preliminary pass on all the boards to read the grain. Then after judging the quality of the surface, mark an arrow on the edge of each to indicate the grain direction, and feed the board into the planer again in the direction of the arrow. After passing the entire lot through the planer twice, put on new blades for the final pass. The expense (and trouble) of changing the blades is well worth it—the last thing you want is patches of tearout or tracks caused by a nicked blade.

Even with fresh blades, however, you'll probably still experience some tearout, but it's nothing to worry about. A little irregularity will lend a more authentic look. In the 18th century, floors were hand planed, then scraped flush and smooth. So a little tearout was unavoidable and expected.

At this point, you might think you're ready to apply a finish, but there are two more things to do: relieve the underside to prevent warp, and round the top edges to help hide irregularities in the gaps between the boards.

Relieving the underside of the boards

Finishing one side of a board and not the other will cause the board to cup. On the face side, the wood will swell as it absorbs the stain and floor finish, but the underside will continue to dry and shrink, causing the board to crown slightly and making it difficult to nail down flat. The wider the board, the greater the risk, and with boards nearly 12 in. across, you can count on this happening.

A technique I have used successfully in flattening antique tabletops is to channel the underside with $\frac{1}{4}$-in. grooves. Channeling also works on floor boards. With three or four $\frac{1}{4}$-in.-wide grooves running halfway through on their underside, the floor boards become somewhat flexible (see the photo below) and can be nailed flat even if they cup. Grooving the underside of the floor boards effectively reduces their width. It's like laying down three narrow boards instead of one wide one, except that on the surface the board is still a single piece. A groove $\frac{7}{16}$ in. deep will not significantly weaken $\frac{7}{8}$-in.-thick floor boards.

Grooves cut into the underside of the floor boards will keep the boards from cupping and reduce the chance of their splitting.

The grooves can be cut on the table saw using a ¼-in. dado blade set to cut about ⁷⁄₁₆ in. deep. Make the first groove about 2¼ in. from the edge, then turn the board around and repeat the operation. Then set the blade about 4½ in. from the board's edge for the next pair of grooves.

If there is a lot of moisture in the floor below (such as a damp basement) and you're still worried about cupping even after grooving the boards, you may want to seal the underside of the floor boards with the floor finish of your choice.

Two edge treatments

Installing a prefinished floor presents some minor problems in the area of edge alignments. First, the edges of adjoining boards may not lie flush with each other because of small discrepancies in thickness and flatness from one board to the next. This condition can be unattractive and possibly dangerous. Second, wide pine floor boards will continue to dry and shrink after installation, and unsightly gaps will open up between the boards. Both of these problems can be dealt with effectively by milling either a 45° chamfer or a ¼-in. radius along the edges.

Manufacturers of modern prefinished flooring chamfer the edges of their flooring. Chamfer-ing hides small gaps and other faults between the boards. It is attractive and accentuates the plank appearance of a floor, but it is not found in antique floors. It looks too crisp and deliberate.

If you are after a more authentic look, consider rounding over the edges. In an old house, the floor boards dry and shrink, and the resulting gaps leave fragile and exposed edges. Over time these sections either break off or wear down, causing them to appear almost rounded. Rounding over the edges of your prefinished floor boards more closely imitates the natural wear and tear on a pine floor, producing a slightly casual, more timeworn effect.

My tool of choice for this job is a router with either a chamfering bit or a ¼-in. roundover bit. Both bits come with a ball-bearing guide attached. Normally the bearing will control the depth of cut and prevent any burning along the edge. This usually works best when the edge the bearing rides along is at least ½ in. thick. If the edge is thinner (as it would be with shiplap flooring), there isn't sufficient bearing surface to guide and support the bit and maintain a uniform cut. So instead of using the ball-bearing guide (which I remove from the bit), I control the cut with a ⁵⁄₁₆-in.-thick fence attached to the router base (see the photo below). And since the chamfer is cut differently on

To rout a chamfer, remove the bit's ball-bearing guide and clamp on a fence, which can be repositioned to cut each side of the shiplapped floor boards.

Don't overdue the chamfer! A ³⁄₁₆-in. cut will be sufficient to produce a handsome joint without excessive space between the boards.

each edge of the board because of the shiplap, I move the fence for each half of the cut.

Before you knock the edges off all the boards, first set the router for the proper depth of cut and take a test cut on a scrap of flooring material. The chamfer or rounding over should not remove more than ³⁄₁₆ in. (see the photo above). Remember that when two boards are placed together, the effect will be doubled, and that additional shrinkage over time will make for an even wider gap between floor boards.

STAINING AND FINISHING

During the 17th and 18th centuries, floors were left raw; as time passed, they acquired the dark color and burnished surface we now associate with antique floors. If you want your new floor to look like a 200-year-old floor, you'll need to use a stain and a clear finish.

If you are prefinishing, now's the time to roll up your sleeves and get to work. If you are installing your boards unfinished, skip to p. 66 and read the section on installation, then do the staining and finishing last.

Oil stains

For prefinished floor work, oil stains are what I recommend to get just the right color on the floor. Oil stains are the easiest to mix and apply. They penetrate minimally into the wood and dry slowly, allowing you time to catch drips, clean up overlap, and pick up any excess. With oil stains, you have the time to move the stain around for more even coverage and a better color.

In selecting a shade, there are many factors to consider. How much light is coming into the room? Will there be rugs on the floor? Does the color have to match a floor in an adjoining room? The best way to arrive at a good choice is to prepare color samples.

Most manufacturers provide color chips that show what each color will look like on a particular wood. This is a good place to start, but remember there is likely to be a difference between the color chips and actual samples of the product. Start by buying a small quantity of one color and test it on a piece of scrap flooring. If the color isn't what you want, move on to another color. Stains can also be mixed together to obtain the perfect shade. If you go this route, take good notes on the composition of each batch, so you can reproduce the mix later on.

All the handling and wiping of the stain and the subsequent rubbing out of the clear coating will tend to remove pigment from the surface and lighten the boards. So it's usually a good idea to settle on a sample color that's a little darker than what you originally had in mind. Using a darker color also tends to minimizes small imperfections and any grain or color differences between boards.

Stain can be applied quickly with a roller. Dip the roller into a tray filled with stain, squeeze any excess from the roller, then roll lightly along each board. Roll on the stain with one hand while cleaning up any drips and runs with a rag in the other. Don't worry about even coverage—the roller will deposit more than enough stain onto the board. Any excess can be wiped around to cover the bare spots and even out the color. A small brush can be used to apply stain to the edges and corners of the shiplap.

After the boards have been stained, lay them flat for 24 hours to dry. Don't skimp on the drying period or the stain will bleed when you apply the clear coating, leaving a blotchy and uneven finish.

Clear coatings

The most durable clear coating for flooring is polyurethane, a clear finish with an amber tint.

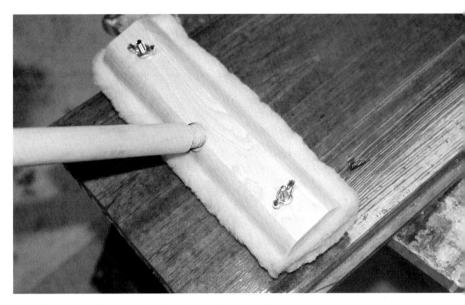

A lamb's-wool applicator can be used to apply polyurethane.

This is what most floor finishers use, applying one coat a day for at least three days. The first coat seals the floor. It is almost completely absorbed by the wood and leaves a splotchy, irregular surface. The second coat establishes an unbroken film and an even color. After it is rubbed out to remove bubbles and bumps, the third coat is applied to flood any knots, gaps, cut-through edges and corners, resulting in a uniform gloss. This is basically the same sequence I follow when prefinishing floor boards.

When working with polyurethane, never shake the contents of the can, which will cause bubbles that can affect your finish. Stir it carefully instead. Pour the coating into a flat paint tray,

and using a lamb's-wool applicator, coat the board (see the photo above). Initially, the surface will be tracked and striated, but as the coating levels and dries, the surface will flatten out.

To ensure a flat and even coating, you can thin the polyurethane by one-third with paint thinner. This will allow the coating to flow and level better, but you may have to apply more coats. As each coat dries, there may be some bubbles or bumps. These can be removed with 120-grit to 150-grit sandpaper or steel wool. For a really smooth floor, buff with steel wool between coats; use 00 between the first and second coat and 000 thereafter. This extra effort will pay off handsomely—your floor boards will absolutely sparkle.

Low-luster finishes

Not everyone wants a floor that gleams. Restorers of old houses looking for a low-luster finish might want to consider Waterlox, a finish that I have used with success. It contains mostly tung oil and looks almost like a penetrating oil with a watery consistency. Since this coating is lighter than polyurethane, it is more forgiving. When applied it leaves no brush marks; bubbles and ridges flatten out and leave an even surface. With the application of additional coats (up to five coats), the floor develops a uniform satin sheen and an evenly filled surface that repels water nicely. I've obtained the best results by applying Waterlox with a foam-rubber applicator or roller and rubbing out with 00 steel wool between coats.

Another advantage Waterlox has over polyurethane is that the floor can be rejuvenated and restored to nearly new condition without professional help. If the floor is damaged, you just repair it, sand, and recoat.

INSTALLATION

With finish applied to all the flooring, it's time to start on the installation. The first step, after removing all the old flooring, is to check the subfloor for soundness. Make sure there are no weak spots, damaged areas, or seams unsupported by a floor joist. The subfloor should be at least ½ in. thick, but ¾ in. would be better. If you encounter any areas that flex, drive 2-in. galvanized screws through the trouble spot into the floor joist. If there is access to the underside of the floor, you may want to squeeze a bead of construction adhesive between the plywood and the joist before driving in the screws.

Layout

Once the subfloor is secured, sweep it clean to remove debris, dried plaster, or joint compound, and to reveal any proud nail heads or fasteners, which should be set. Next determine which direction the flooring will be laid. It should run at 90° to the joists, which is generally the length of the room. Then measure the width of the room and snap a chalkline down the center. This line will be a guide to keep the boards parallel and at 90° to the short walls. Next, snap secondary lines parallel to the center line every 4 ft., then lines indicating the position of the joists (these can be tricky to find in an old house). Since the

nail patterns on your new floor will be visible, you should try to keep them in line with the joists for a neat appearance.

Equipment and tools

The most important tool for installing flooring is the chopsaw. A chopsaw will make a clean, perfectly square cut every time, whether you're quickly chopping an 8-ft. length of flooring in half, shaving off ¹⁄₁₆ in., or sneaking up on a perfect fit. My choice is a 10-in. sliding-arm model. These saws are generally used for trim, but the sliding arm enables them to cut a full 12 in. in width.

Since wide pine boards have a tendency to split, it's a good idea to predrill for nails. Any drill will do, but I suggest at least a 9.6V cordless. It has enough power to drill a couple of hundred small-diameter holes in pine before needing a recharge. The cordless feature means you'll have one less cord to plug or untangle.

Another useful tool is a 2-lb. mason's lump hammer, which looks like a small sledge. This is the perfect tool for driving home the large 10d cut nails used to lay the floor. An old spoon is handy for protecting the wood surrounding the nail heads. No matter how careful you are, an errant hammer blow can horribly disfigure a floor board. And when the board is prefinished,

Here's a tip for nailing down your floor boards without denting them: Make a hole in the middle of an old spoon, and hold it over the nail head as you drive it in.

If you get the first floor boards down square to the room, the rest of the job should go easily. Start by laying a long straight floor board along a wall. Check its position by measuring from a secondary chalk line. Butt one end against the wall and place the other end, freshly cut if necessary to get a clean, crisp butt joint, extending into the room. Now predrill holes for the nails and drive 10d cut nails (rectangular head) through the floor board, subfloor, and into the joists. Leave the heads proud for now, just in case you have to pull up the board later.

Finish the course, butting boards end to end. (In my own work I occasionally use a biscuit joiner to register the ends if an end seam falls on a small hump in the floor.) Then start the second course, staggering the butt ends by at least 16 in. If the seams close up tight, predrill and nail down both courses. Just continue across the room in this manner, periodically measuring the edge of each board from the nearest chalk line to be sure that the boards are running parallel.

If a board will not snug up tight, screw a tapered block to the subfloor about 3 in. from the crooked board (you may need more than one) then carefully drive a wedge between the floor

there's no easy way to fix the damage. A flattened spoon with either a hole in the center or a slot to fit over the nail head will protect the floor as the nail is being driven (see the photo above).

For cutting irregular outlines and cutouts for pipes, a jigsaw is useful. The narrow blades provide good depth of cut and the ability to cut small-diameter curves. Another feature is the

adjustable-angle baseplate, which is useful for undercutting tight profiles.

Nailing down the boards

Since the floor boards are wide and form a graphic pattern on the floor, the butt (square) ends of the boards must be carefully laid out and spaced. Try to avoid placing too many butt ends in traffic paths. Heavy foot traffic could work these butt ends loose over time.

(text continues on p. 71)

JOINTER PLANES

Woodworkers who visit my shop are always surprised to discover that I use jointer planes to prepare the edges of long boards. Most people think jointer planes are relics from the past, suitable only for duty as wall ornaments. They can't imagine how anyone ever used those things to joint and square edges for gluing. At first, I also found it a little hard to believe. The first time I used a jointer plane, it was awkward, a little like using your knees to control the steering wheel of a car while maneuvering around traffic cones.

Jointer planes are used to smooth and flatten the faces and edges of boards. Most early wooden jointer planes are about 24 in. to 30 in. long, with irons 2 in. to 3 in. wide (see the photo below). As with other bench planes, the iron is pitched at 45° and held fast with a wedge.

In the 18th century, jointer planes were commonly used by housewrights, shipwrights, coopers, and cabinetmakers. The one shown here is 32 in. long.

A jointer plane is used to dress and straighten edges and surfaces because its long body can span the small dips on a board while cutting down the high spots. Each time the plane passes over the wood, the two extremes are brought closer together. Eventually the jointer plane will take a single continuous shaving, indicating that the surface is flat. A shorter plane would ride the highs and lows, never flattening the wood, only smoothing it.

I regularly use an old 19th-century jointer plane to dress edges that are too long for my stationary jointer. With a jointer, you run into a problem something like what happens with a short plane, only the wood is moving over the cutting edge. If the bed of the jointer isn't long enough to support the full length of the board, it can never completely straighten it out. With a jointer plane, the material remains stationary and the plane is passed over the wood. The planing can be focused on particular areas with remarkable accuracy; you can address the high spots, while ignoring the lows. This sort of localized planing can't be done on a jointer.

It can take some practice to master this skill of jointing with a jointer plane, but the result is worth the effort.

CHECKING THE PLANE

A plane in poor condition will not work properly, so take a little time to check it out first. Since the jointer plane is made of wood and can warp, the sole must be checked for flatness with a straightedge (refer to the drawing on the facing page as you read this description). If the sole is not flat, true it with another plane or pass it over a jointer.

Next, examine the fit of the wedge. It must hold the iron tight to the bed to maintain consistent projection of the iron and avoid chatter. A wedge should require only a light tap to fix it fast and a lateral snap

Parts of a jointer plane

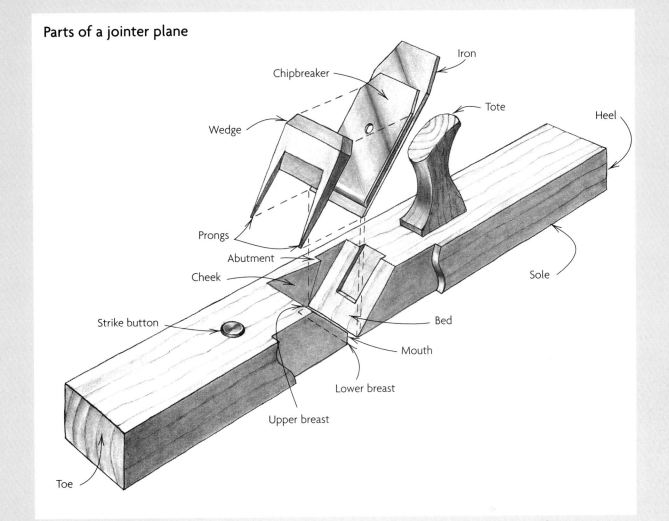

Chipbreaker

Iron

Tote

Heel

Wedge

Prongs

Abutment

Cheek

Strike button

Sole

Bed

Mouth

Lower breast

Upper breast

Toe

to loosen it. Check for any gaps between the iron and the bed. If the iron lies flat without any light passing between the two, then set the wedge in place with a light tap. Look for gaps between the abutment and the prongs of the wedge.

SETTING THE IRON

Once the iron is sharpened it should be set onto the body with the bevel down and the chipbreaker on top. The edge should project evenly across the mouth of the plane by a light $1/32$ inch. While holding the iron assembly in place, replace the wedge and secure it with a few light taps.

If the iron projects more from one side of the mouth than the other, adjust the iron laterally by tapping the opposite side of the iron near its end. To advance the iron for a heavier cut, tap the end of the iron. To retract the iron, hold the iron assembly and wedge firmly, then strike either the heel of the plane or the strike button on the front end of the plane.

USING A JOINTER PLANE

Jointer planes are so large that you might think that they are hard to handle, but that's not really true— controlling the plane is just a matter of becoming

(continued on p. 70)

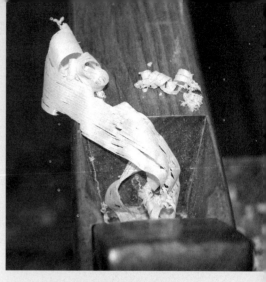

When planing a surface (left), hold the plane firmly at the toe and press the plane to the wood surface for a full chatter-free cut. When planing an edge (above), wrap your hand around the front of the plane with your forefinger pressing against the vertical surface of the board; that will tell you if the plane tilts out of square.

As you plane, check the width of the shaving. When it reaches the full width of the edge, the surface has been planed flat.

(continued from p. 69)
comfortable with the tool. Take it around the block once or twice. Get used to the weight and to the shape of the handle. Become familiar with the plane's movement over wood, across the grain and with the grain. When you're just hanging around the shop, pick up the plane. Get a feel for the tool's balance. These exercises will help you develop a real feel for the tool and get you off on the right foot.

When I plane tabletops or other large surfaces (see the photo above left), I hold the plane by the toe with my left hand and the tote (handle) with my right (I'm right-handed). This position presses the plane flat against the wood surface, ensuring a full cut and a clean surface without chatter marks. If I'm planing an edge (see the photo above middle), my right hand grabs the handle, but my left wraps around the front of the plane with my forefinger pressing against the board. With my hands at both ends of the plane; one on top and the other beneath, I can tell if it rocks even the tiniest bit. I control the beast!

READING THE CUT

Planing an edge square is one of the hardest planing tasks to master. First, the iron must be adjusted for an even projection across the width of the mouth. With that done, the difficulty comes in guiding the plane's bulky body down the length of the edge while at the same time holding it square to the board. With practice this becomes easier, but in the meantime here is a tip that will help.

If a plane held square to the face of a board is guided down an untrue edge, it will produce a shaving narrower than the thickness of the material (see the photo above right). With each subsequent pass, the shavings will become wider. Eventually, the width of the shavings will correspond to the thickness of the board, and at that point the edge will have been planed square. This technique won't guarantee a square edge, but it will work about 90% of the time. You should always check your results with a square.

A few wedges driven along the free edge of a recalcitrant floor board will quickly put it in its place.

board and the block (see the photo above). This should force the floor board into place and hold it there until the nails can be driven in.

Each board should be secured with two nails across its width, and they should be placed about 1½ in. to 2 in. from the edge and spaced at least every 32 in. Occasionally, you might drive additional nails in the center of the board to flatten a stubborn cup. As each nail is hammered into the wood, place the spoon over the head and pound it flush to the floor. This will actually set the nail head just a hair above the floor. You'll see the nail head clearly, but there's no need to countersink it; it won't cause any problems such as

gashing your foot if you walk barefoot or snagging a priceless carpet. The ends of the floor boards should be fixed with nails driven at a slight angle about 1 in. from the end (see the photo at right).

When you're done, take a few minutes to feast your eyes. If you took my advice and prefinished the boards, your new floor looks like a lovingly maintained antique floor right now. You don't have to wait years for great-looking color and a patina to develop. Your prefinished wide-board floor has a soft luster and warm color that would be tough for a floor finisher to match.

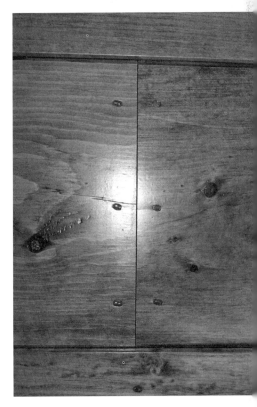

The butt ends of floor boards should be nailed securely to prevent cupping.

6 CHAIR RAIL

Chair rail is probably the least essential trim feature in a room in terms of function. But enter a room with even a simple example in place and something unusual happens. The top of the chair rail is between 28 in. and 34 in. from the floor, so it cuts the wall height into unequal portions. This has the effect of grounding the room and directing the viewer's attention to the lower part. The chair rail effectively creates a waist-high border around the perimeter of a room, providing a focal point on otherwise plain expanses of wall. In a traditional room, it links the windows, doors, and fireplace mantel into a unified design. In a modern home, a chair rail can accent small rooms, foyers, and narrow halls without enormous effort and expense.

Historically, the purpose of chair rail was to protect the wall from the chairs and other furniture placed against it. During the 18th and 19th centuries, the custom was to place furniture around the perimeter of a room when it was not in use. Rooms back then were multi-purpose—a formal front room might be used as a ballroom, or for dining, or even as a bedroom. That meant the furniture was constantly being moved about—within a room or throughout the house.

This reproduction chair rail is simply an elaboration of an early Georgian chair rail. In this example, an astragal molding is laid onto the beaded board, then capped with a half-round. Behind the rail is the foundation strip used to hang it.

Like baseboard (see the sidebar on p. 48), chair rail assisted the plasterer by acting as a stop. Getting a plaster wall smooth and flat was a fairly difficult task. On a wall divided by chair rail, the plasterer had two smaller and more manageable areas to deal with instead of one big one.

Historical chair rail could be plain or elegant. A simple flat board with a bead along each edge might be found in second-floor halls and stairways or in back rooms. In more formal settings, chair rail might be a running frieze of intricate cutouts alternating with short carved flutes and flanked by lively bands of intricate moldings; see the sidebar on pp. 80-81 for one such example.

Chair rail in the early Georgian period (1714-1750) was a large bolection molding. Often it was made up of one or more astragal moldings and capped with a half-round, like the one shown in the photo above. This design created a symmetrical band of moldings in high relief.

Later, chair rails took on a slightly flatter appearance, and the flat space between the top and bottom linear moldings provided a platform for relief decoration. By the Federal period (1780-1820) chair rails were adorned with Greek keys, dentil moldings, and elaborate lines of egg-and-dart carving.

For our chair-rail project I have chosen a design based on a Greek Revival (c. 1830) chair rail I discovered in a small cottage near Cuddebackville, New York. It's a very formal molding, but at the same time playful and lively, with lots of eye-catching shadows. In my house, I used it at the second-floor landing, which leads to the attic. When visitors come to my house and see this molding, they all think I spent a long time carving its clean, fastidious details. "I bet it took you forever to make that stuff," they say admiringly. Really, it was quite easy. What look like hand-carved details details are actually bits of machine-made moldings, cutouts, and shapes, arranged into interesting designs and patterns.

DESIGN AND DIMENSIONS

The chair rail for this project is made up of three parts: the upper molding strip, the decorative band, and the lower molding strip. The band itself has two parts: a decorative face and a backing strip. The entire assembly mounts on a ground strip that is fastened to the wall (see pp. 77-78). For dimensions of the chair rail in this project, see the drawing below.

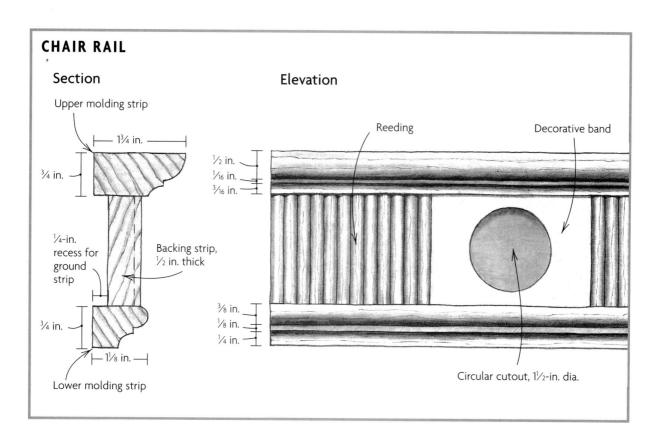

CHAIR RAIL

Section

Upper molding strip

1¾ in.

¾ in.

¼-in. recess for ground strip

Backing strip, ½ in. thick

¾ in.

1⅛ in.

Lower molding strip

Elevation

Reeding

Decorative band

½ in.
1/16 in.
3/16 in.

⅜ in.
⅛ in.
¼ in.

Circular cutout, 1½-in. dia.

If you prefer, you can also design your own chair rail. The only critical dimension is that the molding strips be wide enough to cover the ground strip and project enough to frame the decorative band so it looks good. You can make your own molding strips using common router-bit profiles, or you can get stock moldings at a lumberyard or home center. For the upper molding strip, softer, rounder shapes are easier to sand and shape if necessary; they also fare better in collisions with furniture than moldings with sharp edges. The lower molding strip can be either a smaller version of the upper strip or another profile altogether. There are no hard and fast rules, but to be on the safe side, make up some samples and think about them for a while before going ahead with the whole room.

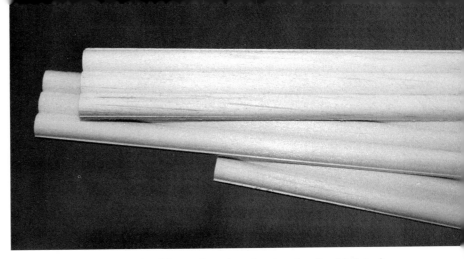

The reeding for the chair rail will be cut from these long lengths of molded stock.

Holes 1½ in. in diameter are drilled on 3-in. centers through a piece of pine, which is then sliced lengthwise into ¼-in. thick strips on a bandsaw.

MAKING THE DECORATIVE BAND

The pattern on my decorative band is composed of a series of circular cutouts with bands of reeding between them. To make the reeds, take a ¾-in.-thick, 36-in.-long clear pine board that's 5½ in. wide and mill both sides using a three-flute router bit. After both sides of the boards are reeded, free the reeds from the board by bandsawing off the sides to ¼-in. thickness, using a fine-tooth blade for the cut. You'll be left with long molded pieces with three reeds running lengthwise (see the photo at top). Repeat as necessary until you have enough reeding for the chair rail you want to make. Then set the reeded pieces aside, and turn your attention to the cutouts.

Begin by laying out several 36-in. lengths of 2-in.-sq. clear pine for 3-in. centers, and using a 1½-in. Forstner bit, drill clear through the pieces. You now have boards with 1½-in.-dia. holes every 3 in. Take these boards and rip them on the bandsaw, using a fine ¼-in. blade into ¼-in.-thick strips (see the photo above).

Cut the reeds to length on the bandsaw, using a notched block of plywood to support the workpiece and ensure a square cut.

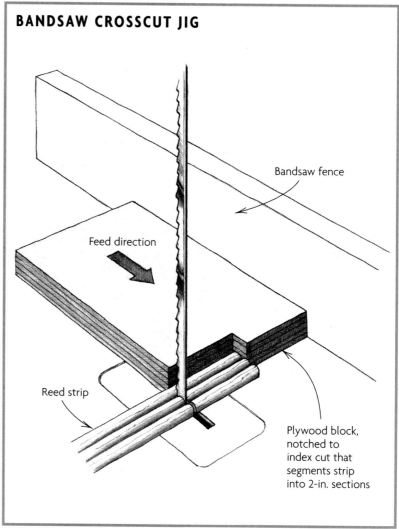

BANDSAW CROSSCUT JIG

Bandsaw fence

Feed direction

Reed strip

Plywood block, notched to index cut that segments strip into 2-in. sections

Once the reeds and circles have been prepared, they have to be cut into segments. The reeds are cut to a length of 2 in., and the circle sections are cut to a length of 3 in., with the circle centered lengthwise. To cut the reeds and the circle cutouts into short sections, use a simple plywood jig on the bandsaw (see the photo above). The jig (see the drawing above) is nothing more than a scrap of plywood with a notch cut out. One end of the jig rides along the bandsaw fence, providing support for the crosscut. The base of the notch is set 2 in. from the blade, and one end of the reeded material is placed against it. The jig serves two purposes: it ensures uniformity of the pieces, and it ensures the safety of the operator. The circle cutout pieces can be cut on a similar jig or on the table saw with a crosscut guide.

To prepare the decorative band, apply yellow glue to a small area of the ½-in.-thick backing strip (a ½-in. thick length of pine or plywood that has been ripped to a width of 2 in). Then lay four

Glue up the decorative band on the backing strip by alternating four sections of reeding with one circular cutout piece, checking occasionally for square.

sections of reeding onto the backing strip, followed by a circular cutout piece (see the photo above), and continue in this manner until the entire backing strip is covered. As you work your way down the strip, clean excess glue from between the reeds and the inside of the circles using a small chisel or a rag.

When you're done, cover the decorative band with a batten, placing paper between the batten and the decorative band, and clamp the entire assembly to a bench top until the glue is dry. Then carefully plane or joint one edge of the decorative band and rip the other edge on the table saw to a width of 2 in.

MILLING THE MOLDING STRIPS

Select a clear ¾-in.-thick, 3-in.-wide length of pine, and on each edge run off the upper and lower molding-strip profiles of your choice. You can use various router bits in combination to achieve the profile shown in the section drawing on p. 74. Then rip the strip to width: 1¾ in. for the upper strip, and the remainder, for the lower strip, to 1 in.

HANGING THE CHAIR RAIL

A while ago I spoke to a friend and fellow woodworker who was hanging a chair rail in a client's

dining room. He complained that the job was impossible. He couldn't find the studs in the wall in order to secure the chair rail. When he got the rail onto the wall, it wouldn't lie flat, and the #8 screws he used to attach the rail to the wall had to be countersunk and plugged. He swore he would never take on another chair-rail job again.

The easiest method for hanging a chair rail is to follow a variation on an 18th-century technique. Back then carpenters would securely nail a ground strip to the studs (see the drawing on p. 78). The wall would then be plastered with a rough coat of plaster, covering the lath strips and ending up flush with the ground strip. Then the chair-rail molding would be tacked to the ground strip with small finish nails, after which the smooth coat of plaster would be applied. Nowadays, there's not much plastering being done, but you can still benefit from using a ground strip.

For ground strips you can use solid-wood scrap thicknessed to ¼ in. or ¼-in. plywood. The strips should be ripped to a width of 1⅞ in.

After you lay out and level the position of the chair rail on the wall, attempt to locate the studs. In stick-frame construction they are spaced about 16 in.

HOW CHAIR RAIL IS ATTACHED

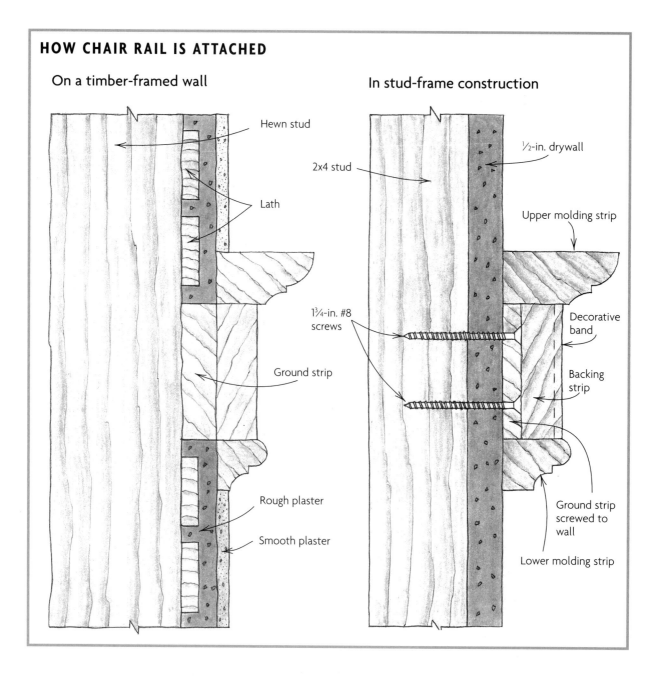

On a timber-framed wall

Hewn stud

Lath

Ground strip

Rough plaster

Smooth plaster

In stud-frame construction

½-in. drywall

2x4 stud

Upper molding strip

1¾-in. #8 screws

Decorative band

Backing strip

Ground strip screwed to wall

Lower molding strip

on center. In older homes, particularly timber-frame structures, where stud spacing was erratic, locating the studs will be a hit-or-miss proposition. Feel free to drill and poke around within

the layout lines, since the chair rail will cover the holes. Attach the ground strip with two screws driven into each stud (see the photo at left on the facing page). It's especially important to

anchor the strips at inside and outside corners. Wherever there is a need for a fastener and a stud cannot be located, use a toggle fastener instead of a screw to secure the ground strip.

Screw the ground strip to the wall with #8 galvanized screws into the studs. If you can't locate the studs (or if there aren't any), use toggle-bolt fasteners instead.

Press the upper molding strip onto the decorative band and against the wall to close any small gaps as you nail it into the ground strip. This small-gauge pneumatic gun uses 1½-in.-long nails and leaves an entry hole the size of a pin head.

With the ground strip in place on the wall, you can plan the layout of the segments. Strive for symmetry and balance, especially at inside and outside corners. If you have to cheat on the pattern, do it on a reeded section. A couple of reeds missing from one section won't be missed.

The decorative band is positioned flush with the top edge of the ground strip and nailed down with small finishing nails spaced every 8 in. to 10 in. Once the decorative band is installed, the lower molding strip can be glued and nailed to its underside. If the molding strip runs out before reaching a corner or a trim casing, the end should be spliced at 45° to another length of molding. At inside corners, the ends of the decorative band can be butted into one another. At outside corners, they should be mitered.

The upper molding strip is the last piece installed. After measuring and cutting the pieces to length, attach them to the upper edge of the decorative band with 1½-in. wire finishing nails or 18-gauge pneumatic nails, spaced 8 in. to 10 in. apart. Now is your chance to close up any small gaps between the decorative band and the wall by pressing the upper molding strip against the wall before nailing it in place (see the photo above right). The chair-rail parts can be primed with a thinned coat of paint before installation. Caulking can be done after installation and before finish painting.

ONE-PIECE CHAIR RAIL

The first step is establishing the rabbet. A moving fillister plane with an adjustable fence and depth stop was used to cut the step.

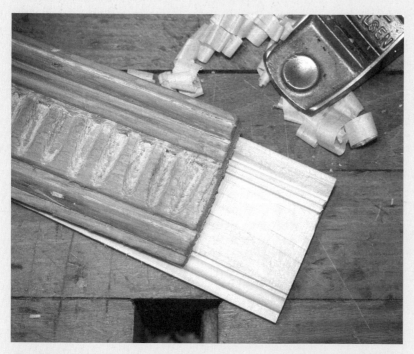

The rabbeted workpiece gets a Grecian ogee with an astragal along the top and a bead with quirk along the bottom edge. Both molding treatments create strong horizontal shadows that draw the eye from side to side.

Over the years I've come across hundreds of chair rails, and most of them were composed of separate parts, like the chair-rail project in this chapter. But the most intriguing are those made from a single piece. Producing a chair rail, or any decorative molding for that matter, from a single piece of wood without using power tools is a demanding task, especially if the design calls for carving. Working with a single piece of wood is risky business. If the craftsman makes a mistake, the entire piece might be ruined or at the very least require time-consuming repairs.

The old chair rail shown in the photos at left dates from about 1810-1830. It caught my eye for two reasons. First, in elevation the molding looks substantial and thick, but in fact it's only $^{11}/_{16}$ in. at its thickest point. Second, even though the molding is small, almost delicate, it makes a big visual impact. The staccato band of cross-grain flutes, flanked by slender linear moldings, takes the eye from one element to the other and back again. This is an exciting piece of work! I was lucky enough to obtain a small length of the original molding. It was muddy-looking and indistinct; all the detail had been flooded with paint and then worn away by caustic strippers and wire brushes.

However, I could still see traces of what it once had been and thought I'd try to reproduce it using 18th-century techniques.

Starting with a clear, straight-grained piece of pine, I cut a large rabbet along the edge of the piece using a fillister plane. The rabbet (see the top photo on the facing page) had to be deep enough to allow a molding plane to cut to full depth along the higher edge with bottoming out. Next, I cut a small rabbet along the lower edge that would allow me to run a small bead about ¼ in. in from the edge (see the bottom photo on the facing page). With the rabbeting done and the molding complete, I began to lay out the carving.

After experimenting with various gouges in order to cut the flutes to the correct width, I then had to fiddle with the spacing of the flutes. If they were too far apart, their impact would be diminished, but if they were too close together I'd run the risk of blowing out the delicate partitions of pine separating them—and remember, these flutes were carved across the grain!

Once I settled on the size of the flutes and the spacing, I began to carve by making the small downward curved cut at the top of the flutes. This was done by pressing the edge of a ¼-in. #3 gouge vertically into the wood about ⅛ in. deep. Next, I selected a gouge with a deep sweep, and beginning at the bottom, took a light, but steady

Small downward-curved cuts along the top mark the position of the flutes and set their spacing. A gouge with a deep sweep cuts the flutes.

The completed molding.

cut, going deeper as I approached the curved end cut. When the gouge reached the end, the thick curved chip simply separated from the board (see the photo at top). Then, with the same gouge, I deepened the cut and corrected

any small error. The completed piece is shown in the photo above. Each flute took a total of three gouge cuts and about 10 seconds to produce.

BUILT-IN WINDOW SEAT

In old houses you often encounter rooms with odd spaces that were created by random alterations and additions. Sometimes a traffic path will cut diagonally through a room, creating an isolated corner. These corners are often too small or odd-shaped for furniture and wind up as wasted space in an already small room. You will find these odd corners and leftover spaces in new homes as well. Try to see them not as flaws, but as opportunities to inject a little character into your home, in the form of a window seat. In the tiny sitting room of my neighbor's Dutch Colonial farmhouse, there was just such a corner to one side of the fireplace and under a window (see the drawing at right). It was the perfect spot for a small window seat.

A window seat can add a lot to a room. In practical terms, it provides comfortable seating and useful storage (which in most houses is sorely lacking). In architectural terms, it visually incorporates forgotten space into the rest of the woodwork in the room.

In this particular room, the space beside the fireplace presented several interesting design and construction problems. Because of a hot-air duct jacketed by a column, access was very restricted. This meant that three corners of the window-seat lid frame would have to be fitted and scribed instead of two. And directly below where the window seat would go, there was another hot-air duct coming up through the floor. This meant that the bottom of the window-seat box would have to be raised high enough to accommodate a flexible hose that could connect the duct under the floor to a decorative grill in the middle of the base molding. On top of all that, the window-seat lid frame would have to be fitted and scribed against the fireplace mantel. The mantel was existing original woodwork, and I didn't want

to diminish its presence in the room by an awkwardly designed piece. So I thought about window seats that I've seen, and considered how best to resolve these problems.

Many period window seats I've examined were simply nailed together. After years of exposure to extremes in temperature, the joints often opened up, allowing dust, moisture, insects, and vermin into the box. Some that were used to store firewood had had their bottom shelf torn loose from the sides by carelessly tossed logs. Others were simply

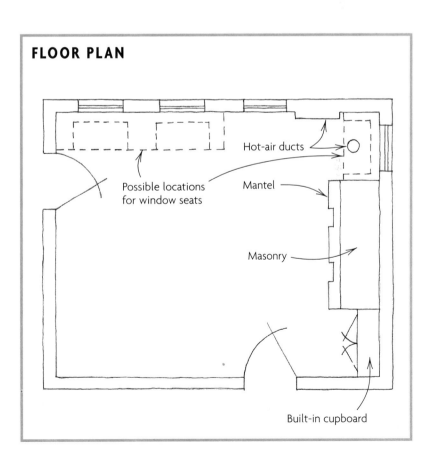

FLOOR PLAN

Possible locations for window seats

Hot-air ducts

Mantel

Masonry

Built-in cupboard

a fascia frame and a lid supported by cleats that were nailed to the wall. The interiors of these window seats were nothing more than exposed floor boards and plastered walls, hidden when the lid was closed. I was after something a bit more elegant.

For this window seat, I decided against a monolithic built-in-place structure and chose instead to build the piece around a box that would be open on top and at the front, with a separate fascia frame and panel to close the box and dress its front (see the drawing below). Con-structing the piece as a separate unit had two advantages. First, it ensured that the panel would be flat and square. Second, because it is separate, the panel could be removed easily and replaced for scribing and fitting without having to struggle with the entire window seat.

EXPLODED VIEW OF WINDOW SEAT

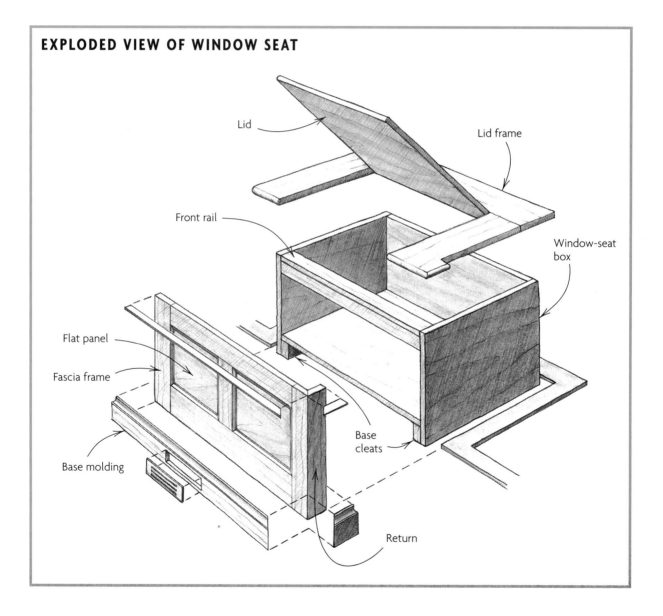

Lid

Lid frame

Front rail

Window-seat box

Flat panel

Fascia frame

Base cleats

Base molding

Return

ESTABLISHING A MODULE

A module is a single unit of measure that appears throughout a room and is used to establish the proportions of other elements in the room. For example, consider the two panels on the window-seat fascia. As you can see in the photo on p. 82, they are similar in design and in size to the panels on the frieze board of the mantel. A relationship has been established, and the pattern carries over from one area to another, creating continuity and harmony.

If you are attempting to design a piece that will integrate well with the existing woodwork in a room, establishing a module can be the key to success. Look around the room for a unit of measurement or an element that repeats itself. It could be a single module or a larger unit that contains a number of smaller modules within it. The drawing below shows how the concept works. In the woodwork at left, the module established by a single pane of glass in the window sash appears again in the four panels of the window seat below. On the door, the module appears within the larger panel. In each instance, the elements might vary slightly in size, but are proportional to the module.

When the principle of modular design is not adhered to, the result is not pleasing. In the drawing, the overall dimensions of the door, window-seat panel, and window on the right are the same as on the left, but they lack a common module. There is no pattern or discipline in the arrangement, and the effect is dissonant.

MODULAR DESIGN

A pleasing balance...
Carefully selecting a module and carrying it throughout the room will produce a calm, harmonious design. In this design, the 8x10 module is used for the window panes and the small door panels. The larger door panels are two modules end to end.

...or an undisciplined jumble?
Without a common module, each element in the wall design stands alone, competing for attention with its neighbors.

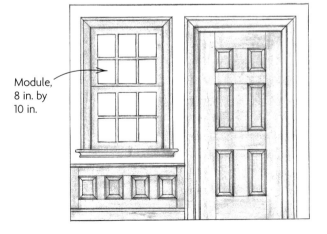

Module, 8 in. by 10 in.

The seat I designed can be fabricated in four sections: the box, the fascia frame, the lid assembly, and the base molding. These sections can be built in the woodshop and then be transported piece by piece to the site. Another advantage is that each part of the window seat can be fitted and installed separately.

As I designed the face frame, I was hoping to match or at least complement the existing fireplace mantel. One of my objectives was to make the window seat look as if it had always been there, and I thought matching some element of the mantel would produce a plausible design. My neighbor and I had briefly considered flanking the window seat with fluted columns, similar to the pilasters on the mantel, or possibly a chip-carved ornament in each of the window-seat panels. But these devices would have mimicked the mantel and detracted from its easy command of the room. We decided a better way was to copy the module of the flat panels on the mantel foundation (see the sidebar on p. 85). This would link the window seat to the mantel in an understated way that wouldn't compete with it.

BUILDING THE BOX

The box won't be seen in the finished piece; its purpose is storage and support. Therefore you want to produce a tight, clean, and durable box and do it quickly and inexpensively, so you can save your skills and energy for the parts that will show.

For my window-seat box, I used ¾-in. shop-grade birch plywood. It isn't as perfect as cabinet-grade plywood, so it's a good choice for the insides of cabinets, for shelving, and other unseen components. It has a smooth surface with a mild grain, it is strong, it is easy to sand and prepare for

finishing, and it takes paint beautifully. Since plywood comes in 4x8 sheets, you can just cut panels to whatever size you want, eliminating the need to glue up narrow boards.

The basic box has an open front and top (see the photo below), with the bottom supported by base cleats of scrap plywood or solid pine. The sides and bottom are butted and screwed, then the back is attached onto the H-shaped form, closing in the box and providing stability. A rail spanning the front of the box provides additional strength and a ground for the mounting of the fascia frame and support

The box of the window seat is a simple plywood H-form with a full back. It's quick and easy to build, and makes a tight, clean compartment for storage.

for the lid frame (see the drawing on p. 88). It's made of two 3-in.-wide pieces of ¾-in. pine joined at right angles, like an L.

Dimensions

Build the box about 3 in. narrower and shallower than the opening, leaving room for the cleats that will be nailed or screwed along the walls of the opening and allowing the outside of the box to be shifted slightly from side to side, if necessary, when it is installed (see the drawing on p. 91). Window seats are generally 17 in. to 18 in. up from the floor, including cushions. Depth can run from 16 in. to 24 in., providing anything from a narrow perch to a luxurious throne. Length can run from the width of a window to the full length of a wall.

Assembling the box

To hold the box parts together at right angles, use Universal clamps (see the photo above right). These clamps are better than a second pair of hands and have the added advantage of eliminating the bowing and distortion caused by pipe clamps. Four to six Universal clamps will hold the entire box together until the parts are secured with screws.

With a combination tapered drill and countersink bit, drill a series of holes approximately 6 in. apart through the sides and

Universal clamps hold the box parts rigid and square, making it easy to drive the screws. Unlike antique nails, galvanized screws will keep this joint tight for centuries.

into the bottom piece. Drill and countersink bits produce a perfect hole in any material without the risk of splitting. Then drive Phillips-head #8 1½-in. galvanized screws to just below the surface of the plywood. This joint will hold forever and even resist hammer blows to knock it apart, yet be easy to dismantle should the need arise, as it did for me. The back of the box is screwed to the H assembly.

I built my window-seat box 3 in. narrower than the wall opening, but that wasn't enough to clear the jog in the wall that housed the air duct. There was no way I could tilt, squeeze, flex, or cajole the box into the hole. After countless attempts to shoehorn

the thing into place, I gave up. I had no recourse but to dismantle the box, shorten the bottom, the front rail, and the back by l in., and then reassemble it. Well, the whole operation took me less than 15 minutes! The speed and ease of the alteration were due to the simple design of the box— no fancy joints, no glue, no delicate parts.

Screwed to the sides of the box are full-length cleats, running front to back along the bottom, supporting the bottom of the box (in case it's used to hold firewood) and providing additional ground for the installation of the face frame and base molding. You don't have to attach the base cleats before you clamp and screw on the plywood parts, but if you do, they will provide support. In this particular window seat, the cleats serve another purpose as well: they raise the bottom up off the floor enough for the flexible hose that redirects the heat from the floor vent.

Finally, attach the front rail to the assembled plywood box with three screws at each end. Like the bottom, it is joined to the sides with countersunk Phillips-head #8 galvanized screws (see the drawing below).

BUILDING THE FASCIA

The fascia for the window box consists of three stiles, a top and bottom rail, and two flat panels. The left and right stiles are full height, extending from the floor to the top of the plywood box. The right stile is 3 in. wide, with a mitered return that tucks behind the fireplace mantel because that side of the window seat projects beyond the mantel. The left stile is 4 in. wide; the extra inch allows for scribing to the wall. The rails are 3 in. wide. The medial stile, which connects the top and bottom rails and separates the panels, is 3 in. wide.

Its length was determined by measuring from the top and bottom rails once they were situated along the length of the outer stiles. Stiles and rails are a clear section of #2 pine.

The panels can be made of either plywood or solid wood—it doesn't matter. I trolled my scrap lumber rack and came up with two ½-in.-thick solid pine boards that were perfect.

Since the back of the panel will not be seen, a rabbet cut on all four sides works fine to join the panel to the frame. Rabbeted edges are thin enough to negotiate the frame, yet most of the panel's thickness will remain. The rabbet can be cut on a table saw or with a router. If you prefer, you can simply bevel the panel all around with a block plane until if fits into the frame groove.

On this window seat, I wanted a flat panel that would match those on the nearby fireplace mantel, so I reversed the normal orientation and put the raised surface toward the back of the frame. A thin ¼-in. panel would have been easier to work with, but too fragile in this application. A window seat that is used for storage and seating often receives very rough treatment.

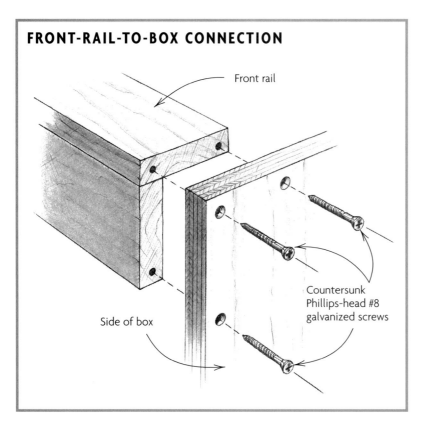

FRONT-RAIL-TO-BOX CONNECTION

Front rail

Side of box

Countersunk Phillips-head #8 galvanized screws

All framing members are molded on the edges that hold the panels (see the drawing at right). Both outer stiles are molded on their inside edges; the outside edges are left square. The top and bottom rails are molded on their inside edges and their ends. The medial stile is molded on both edges and ends.

The best way to make the molded cuts is to use cope-and-stick router bits. These bits are sold in matched sets (see the top photo on p. 90; one cuts a small quarter-round molding and the groove on the edge of the material and the other makes the coped cut on the end. Although the resulting joint isn't a true mortise and tenon (see the bottom photo on p. 90), it fits snugly and offers enough of a glue surface to serve as a good substitute. And because the joint is simply coped and not mortised into place, you can slide the coped end along the molded edge of the stile for the precise arrangement you want. Then measure the areas between the frame members for the panels, and on all sides add an amount equal to the depth of the groove to ensure full engagement of all parts. If you use solid-wood panels, a small allowance should be made to accommodate their movement within the frame. If you use plywood for your panels, no adjustment is necessary.

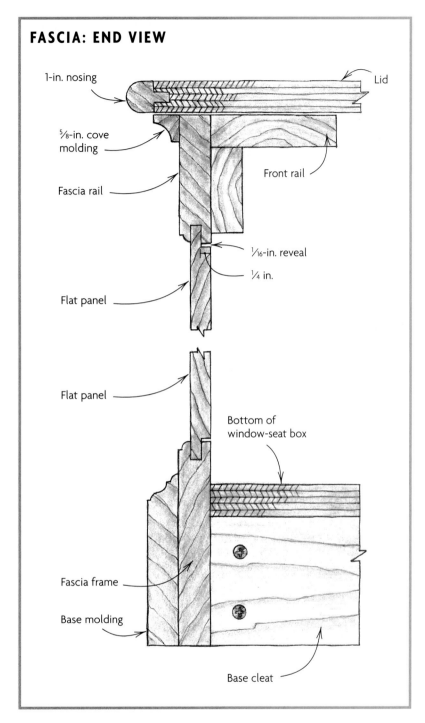

FASCIA: END VIEW

1-in. nosing

⅝-in. cove molding

Fascia rail

Flat panel

Lid

Front rail

$\frac{1}{16}$-in. reveal

¼ in.

Flat panel

Bottom of window-seat box

Fascia frame

Base molding

Base cleat

Cope-and-stick router bits (right) produce matched parts for frame-and-panel work. A cope-and-stick joint (below) isn't a true mortise-and-tenon joint, but it fits and holds well enough to serve as a good substitute.

With the box leveled, mark its top edge along the walls; this line will locate the cleats that will secure it in place (see the drawing on the facing page). Now remove the box and install the cleats. The 1x2 cleats are screwed with fasteners every 12 in.; that way you are sure of hitting a stud or other structural member in the wall. Then reposition the box, shim if necessary, and place #8 1½-in. screws through the sides and back and at the corners where they won't be seen when the lid is opened. The top of the cleats will also support the lid frame (see pp. 93-94).

INSTALLING THE FASCIA

Once the window-seat box is screwed to the cleats, you can install the fascia frame and panel. In many installations, the fascia frame will extend across the entire recess, reaching both side walls. The window seat I built extended beyond the fireplace mantel, so one corner projected into the room. That meant that the leading corner of the fascia frame needed a return into the mantel. This is simply a right-angle extension of the fascia. For a clean appearance, I mitered, glued, and nailed the edges of the right-hand stile and the return.

INSTALLING THE BOX

Set the box into the wall recess and check it for level. If you determine that the box is a little too high on one side, you can either slip a shim under the other side to raise it or cut down the high side. Whichever method you choose, the fix will be covered later by the base molding, so don't agonize over it. You should repeat this process front to back also. What do you do if the floor and window sill are significantly out of level? Do you match existing horizontal surfaces or go for true level? Generally, you should set everything level, since people have the uncanny ability to notice things that are out of true.

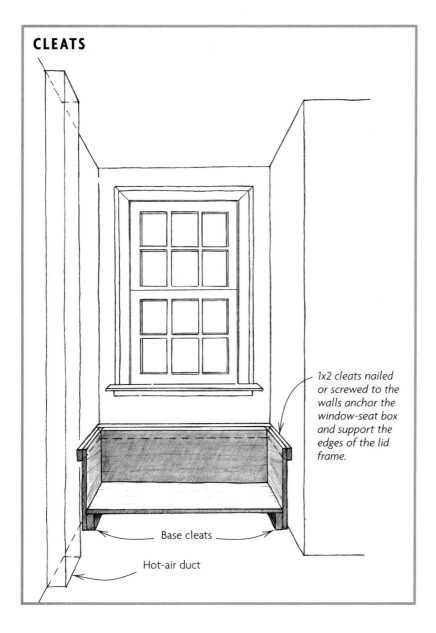

CLEATS

1x2 cleats nailed or screwed to the walls anchor the window-seat box and support the edges of the lid frame.

Base cleats

Hot-air duct

it off. If the blade sits away from the wall at the bottom, take an amount equal to that gap from the top. When one side of the fascia frame fits tight to the wall, repeat the procedure on the other side. Your cuts should leave the panel slightly over-sized, so you can plane it to fit the opening exactly.

The fascia frame is attached by drilling and screwing through the back of the front rail and into the back of the fascia frame. For this, use #8 1-in. screws, spaced about 14 in. to 18 in. apart. At the bottom of the panel, you can place counter-sunk screws through the fascia stiles and into the base cleats; these screws will be hidden by the base molding.

BUILDING THE TOP

The top of the window seat has two parts: a frame and a hinged lid. The frame, which will be fitted and scribed to the opening, is made of ¾-in. pine. The lid is made of ¾-in. plywood, edged on three sides with a 1-in. nosing (see the photo on p. 92). Using plywood circumvents the shrink-age and warping problems that you would encounter with a solid-wood lid.

Because the walls flanking the window seat are unlikely to be plumb, the first order of business is to match their slopes on the end of the fascia. You can use a framing square to deter-mine how much has to be taken off the fascia frame. Lay the short blade of the square on the top edge of the window-seat box and place the long blade against the wall. If the blade sits away from the wall at the top, take an amount equal to that gap from the bottom of the fascia frame, using a handsaw or jigsaw. If the amount is small (¹⁄₁₆ in. or less), you might consider planing

A ¼-in. edge of solid pine tongued into the edge of the plywood lid protects and hides the plywood layers. It can be shaped into a radiused nosing with a router.

Building the frame

The frame can be built in a U shape, with one back rail and two side rails, or fully enclosed, with a front rail, a back rail, and two side rails (see the drawing on the facing page). The U-shaped frame is easier to build and provides easier access to the contents of the window seat. The only drawback of this design is aesthetic: From the front of the window seat, the line of the lid is broken by the separation between the frame and the hinged lid. The fully enclosed frame can be used if you want to disguise the storage compartment.

To build the U-shaped frame, begin by measuring the opening from wall to wall (corner to corner). Lay out the width and length of the frame members and their projection over the front edge of the window-seat box.

Take into account the thickness of the fascia frame and panel, any small moldings you may add to the top of the fascia, and the eventual profile of the front edge.

The inside of the frame must be square to accept the plywood lid, and the outside must be scribed and cut to fit tightly against the wall. Since you can almost count on the wall bowing or the corners being out of square, getting the frame to fit well will be a difficult task at best. To make it a little easier, measure and scribe one frame member at a time, starting with the back rail. Set an oversized piece in position against the wall. Make sure the inside edge of the rail is parallel to the front of the window-seat box. Now locate the widest gap between the back edge of the rail and the wall and set the legs of your compass to it. Keeping the compass upright, move the pin

leg along the wall and the pencil leg along the surface of the rail. This will give you an outline on the rail that mirrors the contours of the wall. With a jigsaw or a coping saw, cut to the outside of the scribed line. Now place the scribed edge against the wall; it should fit well. Next, determine how wide the back rail will be and cut any extra material off the width. When the back rail fits, repeat the scribing process for the side rails.

Since the frame will be supported by the edge of the window-seat box and by the cleats attached to the wall, there is no need for super-strong joinery; biscuit joinery or a shallow tongue and groove will suffice. Make sure to dry-fit the frame and place it in position before you glue it up.

Building the lid

Plywood makes a strong panel, but its edges aren't pretty to look at. For most lids, I recommend gluing a ¾-in. solid-wood nosing onto the edges of the panel. At the very least, use a solid nosing on the front edge and thinner veneer edging on the sides and back. In sizing your panel, you need to subtract the width of the edging from the overall dimensions of the opening and also allow a generous ¹⁄₁₆-in. reveal all around for clearance.

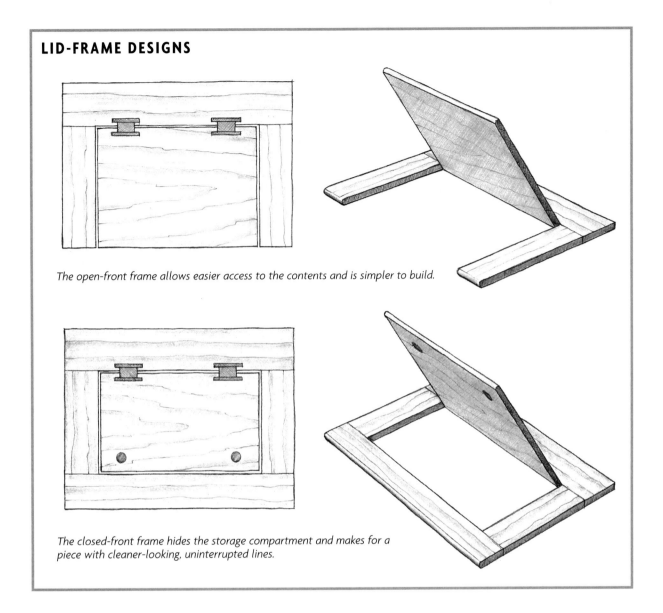

LID-FRAME DESIGNS

The open-front frame allows easier access to the contents and is simpler to build.

The closed-front frame hides the storage compartment and makes for a piece with cleaner-looking, uninterrupted lines.

INSTALLING THE FRAME AND LID

When the frame is dry, set it into place and check the fit. Shave down any tight spots with a block plane or spokeshave until the frame drops into place and lies flat. Next, check the inside of the frame for square and measure the opening to ensure an even reveal all around the lid.

Now the frame can be attached to the cleats or to the edge of the box. The best way is to shoot #8 1½-in. screws into countersunk holes spaced about 6 in. apart.

When the frame is secured, fill any gaps between the frame and the wall with shims. Fit the lid into the opening to double-check the evenness of the reveal, and center the hinge knuckle over the reveal. Mark the lid and the back rail of the frame for screw holes. After drilling the holes, secure the hinges with pan-head screws.

If you like, you can install the lid without any hardware at all. I have seen several drop-in lids that are supported by a small ledge on the underside of the lid frame.

Once the lid is installed, you'll need a way to lift it up. With the open-front design, the lid can be lifted by its overhanging edge. The drop-in seat needs a 1¼-in.-dia. hole about 3 in. from each front corner of the lid, with the top and bottom edges of the hole chamfered with a router (see the photo on the facing page). Another solution is to install a flush pull. A flush pull is a plate with a ring pull recessed into it. These are listed as nautical fittings or campaign hardware in mail-order catalogs.

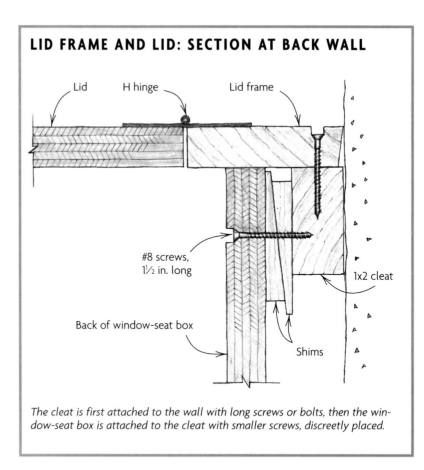

LID FRAME AND LID: SECTION AT BACK WALL

Lid   H hinge   Lid frame

#8 screws, 1½ in. long

Back of window-seat box

1x2 cleat

Shims

The cleat is first attached to the wall with long screws or bolts, then the window-seat box is attached to the cleat with smaller screws, discreetly placed.

THE BASE MOLDING

On a piece such as this, designs for base molding can vary. Generally the base molding is ¾-in.-thick solid wood. It should be at least as high as any surrounding molding. Often I make it about 1 in. higher than the room's base molding. That way the existing molding can be cut at 90° and die into the window-seat base molding without any coping or mitering. This produces a clean junction and gives a little prominence to the window-seat molding. The molding can be attached with counter-sunk 4d finishing nails or a pneumatic nail gun.

THE HOT-AIR VENT

Since hot-air vents are normally situated near windows and doors, accommodating a vent into a window-seat design is a common problem. The task is accomplished by extending the vent by means of a flexible hose, which links to a metal conversion boot that fits into the back of the base molding. A trim grill goes on the front of the molding. All the fitting-to-hose connections should be secured with a hose clamp and duct tape.

FINISHING TOUCHES

As you design and build a piece such as this, one of your objectives should be to have the new work fit seamlessly into the existing woodwork in the room. Too often new work, no matter how well executed, will be recognized as such. If you examine old

At the front corner of the lid, 1¼-in.-dia. holes serve as a finger grips. The perimeter of the holes should be softened with a roundover bit.

woodwork, one of the first things you notice is that edges and reveals are softened by wear and corners are filled in by paint. Profiles that were once sharp are now slightly blurred, hinting at bygone decades and sometimes centuries. Old woodwork evokes a sense of history that has tremendous appeal.

The easiest way to give your new woodwork an instant past is to break sharp outside corners with sandpaper and fill in seams and inside corners with caulk before you paint it. Caulking will give the appearance of a surface that has been painted several times. Woodworkers often hesitate to use caulk because they think it will destroy the sharpness of their molding profiles and

detract from all their hard work. Caulk simply softens hard lines and eliminates distracting shadows. It actually improves the appearance of new work— and helps to blend it in with existing woodwork.

Use a high-quality water-based paintable latex caulk. For the most efficient application, use caulk in a tube container that is applied with a gun. The tube is loaded into the barrel of the gun and the plunger is pressed against the end of the tube. Squeezing the trigger of the gun advances the plunger against the end of the tube, forcing caulking out of the pointed end. When you want to stop the flow of caulking, relieve the plunger pressure by disengaging the

teeth on the underside of the plunger with a quick 90° turn or the caulk will continue to flow from the tip of the tube, spilling out all over the place.

I apply caulk on all seams to soften inside corners and lightly blur joints. Only the smallest bead of caulk should be applied to a seam or in a corner. If more is needed or desired, apply a second bead on top of or alongside the first bead. Once the entire seam is caulked, the material should be smoothed out and feathered with a clean fingertip, followed by a damp sponge. No excessive buildup or smears of caulk should remain on the woodwork. Any excess should be immediately cleaned up with the sponge before the caulk dries.

Formerly, hinges were painted the color of the surrounding woodwork, but today many hinges come with a slick black paint that retards the adhesion of additional paint. In order for your final color to adhere properly to the hinges, you should remove the factory-applied finish. This can be done with a propane torch or by placing the hinge over a lit gas burner on a kitchen stove. Once the metal changes color, quench it quickly in water. This treatment will produce an attractive mottled gray surface that will take and hold any interior latex or oil-based paint.

8 FEDERAL FIREPLACE MANTEL

Whenever I enter the parlor or front room of a period house, I immediately look for a fireplace. The presence of a fireplace tells me of the room's importance in the home life of the family that first lived in the house. The quality and detail of the mantel's woodwork tell me about the family's position in society and about the maker as well—his ability as a craftsman, his training and background,

and his tools. Since the mantel was purely decorative, it often served as a display piece for the craftsman's skill. Everything he knew would go into it—the trickiest moldings in fantastic combinations, elaborate carved designs, and turnings. It often turned into a tour de force.

A mantel is a simple architectural structure composed of a roof (the shelf) and supporting

foundation (the pilasters and rails). Everything else—frieze bands, panels, cornice moldings—is ornamental. The purpose, of course, is to frame the fireplace opening and draw the eye to the hearth.

The mantel featured in this chapter was designed for a small New York City apartment; it is shown to scale in the drawing below. To adapt the design

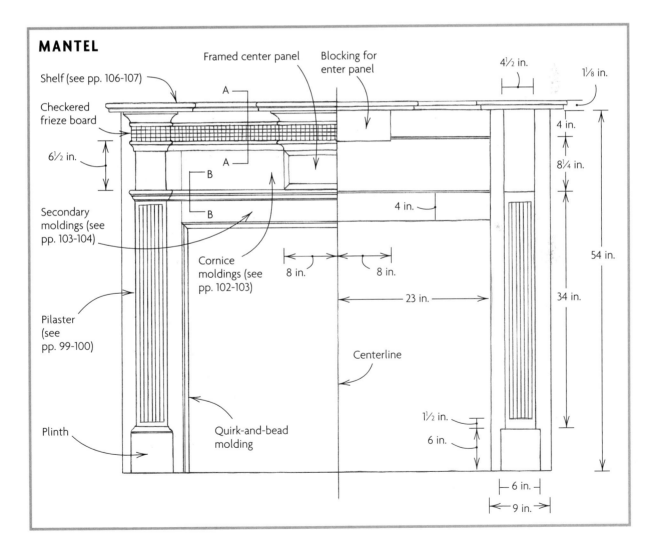

MANTEL

Shelf (see pp. 106-107)

Checkered frieze board

6½ in.

Secondary moldings (see pp. 103-104)

Pilaster (see pp. 99-100)

Plinth

Framed center panel

Blocking for enter panel

A

A

B

B

Cornice moldings (see pp. 102-103)

Quirk-and-bead molding

4½ in.

1⅛ in.

4 in.

8¼ in.

4 in.

8 in.   8 in.

23 in.

Centerline

54 in.

34 in.

1½ in.

6 in.

6 in.

9 in.

successfully to your own house, you'll need a fireplace with an opening from 40 in. to 48 in. wide, with the lintel (the iron piece that holds up the brick above the front of the opening) between 32 in. and 36 in. above the hearth. A fireplace much larger or smaller could distort the proportions, so you might want to rethink the design.

BUILDING THE FOUNDATION

Lay out your foundation to expose about 8 in. of brick—about the length of one brick—inside the opening. Exposing the brick introduces another material against the mantel, creating an interesting contrast in texture. It serves a practical purpose as well, by protecting the mantel from fire or discoloration caused by smoke blowing back into the room. The exposure of the brick facing doesn't have to be uniform if it will create an ungainly mantel. Play with it a little. Try to end up with a mantel of graceful and sensible proportions with a shelf height of 52 in. to 60 in. above the hearth.

The material for the foundation must be thick enough to provide a rigid surface for the pilasters and other ornamentation. You can use #2 grade ¾-in.-thick pine if you can cut out the knots or

plan your layout so they will be covered with pilasters and moldings. If the mantel is to be large, consider making the foundation out of 5/4 stock.

The vertical sections of the foundation can be made of a single board or for the sake of economy, glued up from narrow strips. If the section is glued up, be sure that any seams will eventually be hidden by the pilasters.

One of the most common problems with old mantels is a tendency for the wide horizontal rail to crack along the grain. This usually happened because no allowance was made for the inevitable shrinkage across the grain of a wide piece of wood. A fireplace generates a great deal of heat, even through thick masonry walls, and this heat would often cause the rail to shrink, resulting in long,

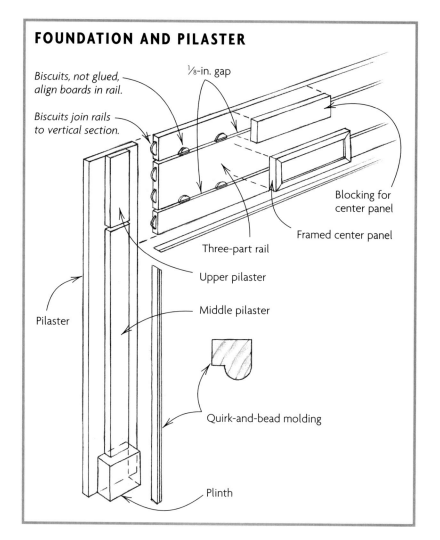

FOUNDATION AND PILASTER

Biscuits, not glued, align boards in rail.

Biscuits join rails to vertical section.

⅛-in. gap

Blocking for center panel

Framed center panel

Three-part rail

Upper pilaster

Middle pilaster

Pilaster

Quirk-and-bead molding

Plinth

The rails and vertical sections of the foundation are joined with biscuits, glued, and clamped. In the expansion joints between rails, unglued biscuits maintain alignment, and spacers ensure a uniform ⅛-in. gap.

meandering cracks that disfigured otherwise flawless woodwork. The foundation design in this project avoids the problem by using three narrower boards with ⅛-in. gaps between them, which serve as expansion joints (see the drawing on the facing page). In the finished mantel, these expansion joints are concealed by moldings.

On period mantels, I have often seen long mortise-and-tenon joints used to join the vertical sections to the horizontal rail of the foundation. Since this is a small mantel and to speed up the work without sacrificing strength, biscuit joints can be used instead. On the rail-to-vertical section, space the biscuits no more than 4 in. apart.

Biscuits should also be used to register the three parts of the horizontal rail, but don't glue them in (if you did, the parts of the rail wouldn't be able to expand and contract freely). Space the biscuits about 6 in. apart.

Use yellow glue to coat the joints and biscuits on the rail-to-vertical-section joints, then clamp the assembly with pipe clamps). On the expansion joints, place ⅛-in. spacers between the parts of the rail to ensure an even gap (see the photo above right). When all is

ready, check for flatness across the foundation with a straightedge. Also make sure that the opening is square.

ADDING DECORATIVE ELEMENTS

Once the foundation is complete, set it aside while you work on the various decorative elements that will complete the mantel—pilasters, frieze band, central panel, cornice moldings, and secondary molding. These can be built in any order that suits you; let's start here with the pilasters.

Pilasters
Pilasters are the vertical elements that support the shelf. They are divided into three sections: plinth, middle, and upper (see the drawing on the facing page).

Plinth The plinth acts as a base and anchors the design. It is usually thicker and wider than the middle section and is often scribed to the floor or hearth. Plinths should be made of 1½-in.-thick wood and dimensioned to allow a ¾-in.-wide shelf around the middle section for a small molding.

Middle section The pilaster's middle section is one of the mantel's principal decorative elements. It can be left plain and be framed with a small molding or filled with flutes or reeds. Reeds lend a strong vertical direction that is visually stronger and more energetic than flutes. Each reed casts a shadow that makes the pilasters look more substantial. The simplest way to make the reeds is to cut them with a router bit on a router table. Look for a reeding bit that cuts the beads close together, resembling the work of a molding plane (cheaper router bits have wide grooves between the beads). I use a bit that cuts three ¼-in.-dia. reeds at a time (it's the same bit I used to cut the reeds for the chair rail in Chapter 6).

To make the middle section, start with a clean pine board ¾ in. thick and about 6 in. wide (small knots are acceptable, but the board should be otherwise unblemished). Mill the reeds on both edges of the board, cut them free from the board on the bandsaw (see the top photo on p. 75), then repeat the process until you have enough three-reed strips to cover both pilasters. After planing and cleaning up the edges of the reeded strips, glue and clamp them down onto the pilasters.

When the assembly is dry, frame the reeded area with pine strips (see the photo below). These strips give the reeded panels a more finished appearance. Since the ends of the pilaster middle sections will be covered with moldings, you can fasten the frame strips to each other with countersunk screws at the corners. The frame members can be attached to the pilaster with nails or narrow-crown staples.

Upper section The upper section of the pilaster is left plain and positioned as a continuation of the middle section. This creates the impression that the reeds of the middle section were carved and stopped into a single piece of pine. This section of the pilaster is screwed directly to the foundation. The screws should be placed where they will eventually be covered by the cornice and narrow frieze moldings.

Subassembly Center the pilasters on the vertical section of the foundation and glue them to the foundation board. When the glue has dried, screw them in place from the back of the foundation.

The plinth blocks are first scribed to the floor or hearth, then attached to the foundation with screws through the back. If this sequence is not convenient, the blocks can be countersunk and screwed from the front, then plugged.

Checkered frieze band

Woodworkers in the 18th century often tried to make things look more complicated than they actually are, and the checkered frieze is a good example of a decorative detail that at first glance boggles the mind. Everyone who

The frame around the reeded pilaster is screwed together with #8 screws, which in the finished mantel will be covered with molding.

sees the checkered relief on this mantel thinks I spent weeks carving out alternating squares! I don't have that kind of time or patience. In fact, this eye-catching molding treatment, which I discovered on an old Federal doorway that was being dismantled, revealed its secret to me as soon as I turned it over. The craftsman had cut a series of dadoes in a wide piece, ripped the piece into narrow strips, and then offset the strips to create the checkered pattern. Nowadays, of course, modern machines make this operation a snap.

The checkered frieze in this project calls for a ½-in., four-course checkered border ⅜ in. deep, so you need to make an indexing jig that spaces these ½-in. cuts exactly ½ in. apart. (This type of jig is more commonly used for cutting finger joints, and for this operation it works much the same way.) Begin by setting up table-saw dado blades for a ½-in.-wide cut ⅜ in. deep. Clamp a wooden fence to the miter guide, and cut a dado through the fence. Move the fence to the right precisely ½ in., screw it to the miter guide, and cut a second dado. With the fence fixed, set an indexing pin into the first dado on the fence (see the photo at top right). Make the pin ¹⁄₁₆ in. lower than the slot to be sure that the pin won't lift the workpiece off the saw table.

A finger-joint-style jig made for ½-in.-wide dadoes exactly ½ in. apart is the secret to the checkered frieze. For this design, set the height of the dado blades to ⅜ in.

For the first dado, place the end of the workpiece against the stop.

For the frieze, you will need to rip a board of ¾-in.-thick clear pine about 3 ft. long to a width of 3 in. (this allows for four ½-in. courses, plus an extra course and another ½-in. allowance for saw kerfs). Place the board's edge against the miter-gauge fence (with jig attached) and the end against the indexing pin (see the photo above). With the dado blades set for a ⅜-in.-deep cut, make the first dado. Then set the cut dado over the indexing pin and cut the second dado. Continue in this manner, moving the board over a notch after each cut, until you reach the end.

Next, rip the dadoed board into strips ½ in. wide. In order to produce the checkered effect, simply offset the strips (see the photo below) and glue them together. When the assembly is dry, carefully plane the back side flat and glue it onto a ½-in.-thick backing strip for support.

The next step is to resaw the checkered lamination to final thickness since some surface misalignment of the squares on the checkered band is unavoidable. A bandsaw fitted with a fine-tooth blade will do a good job. If you don't have a bandsaw, a table saw can be used, but the cut should be made slowly in order to avoid chipping the checkered band. After resawing, the recessed portions should be about ³⁄₁₆ inch deep. Finally, remove any saw marks from the face of the band with a block plane or sander.

Central panel

A central panel breaks up the expanse of the frieze board and adds visual interest. Along its bottom edge, it rests on a narrow frieze molding. The panel's top edge is wrapped by the cornice moldings.

The central panel is made up of two pieces of pine. The upper one serves as blocking for the wraparound cornice moldings (see the following section); the lower one, which is visible, is made up of a ⅝-in.-thick flat field surrounded by a ¾-in.-wide mitered frame (see the drawing on p. 98). Since the field is less than 6 in. wide, you don't have to allow for significant cross-grain movement. Simply glue and nail the frame molding around the flat panel. I thought about decorating the panel, but decided against it so as not to compete with the checkered molding, which is the main decorative element. When the central panel is completed it can be attached to the foundation by nailing it through the front or with screws from the back.

Cornice moldings

The owners of the apartment where this mantel was to be installed told me they enjoyed entertaining by candlelight, so I chose a molding arrangement that would produce dramatic shadows in soft light. Although the moldings are unified and seem to flow into one another, they are actually five separate moldings: the shelf, the large cove, the upper bead, the lower bead, and the small cove (see the drawing on the facing page). For the mantel to succeed all the parts have to relate proportionately, and the viewer's eye must move smoothly from one element to the next.

After the 3-in. board is dadoed, rip it into ½-in.-wide strips. Offsetting the strips creates the checkerboard effect.

CORNICE MOLDING (A-A from p. 97)

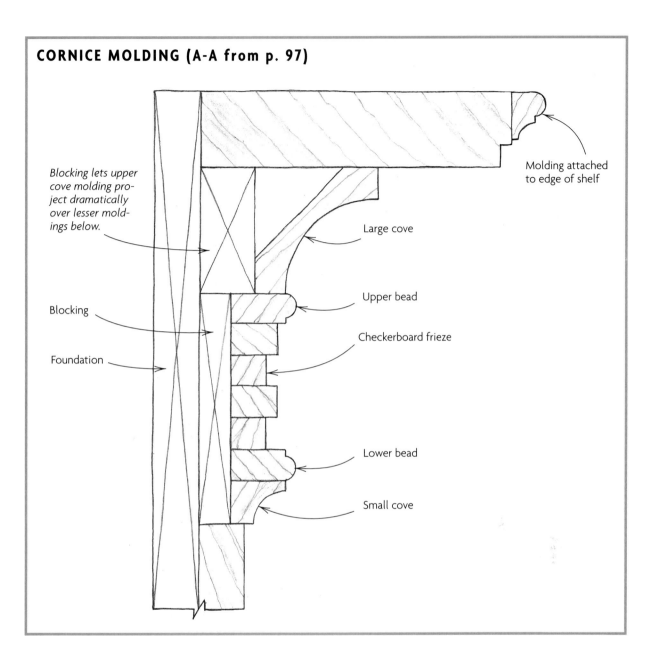

Blocking lets upper cove molding project dramatically over lesser moldings below.

Molding attached to edge of shelf

Large cove

Upper bead

Checkerboard frieze

Blocking

Foundation

Lower bead

Small cove

On this mantel, a large sweeping cove leads from the top of the pilasters to the underside of the mantel shelf. The smooth inverted curve creates a nice transition and gives the mantel lightness and vertical movement. Below that, a full half-round bead contrasts nicely with the sharp squares of the checkered band. Along the bottom edge of the checkered band, the bead repeats, for balance. Under the lower bead, a small cove molding lends a little more lift.

Other moldings

Below the cornice moldings is a secondary molding—a small band topped with a ½-in. beaded shelf (see the drawing on p. 104). This molding performs two functions. First, it establishes the lower edge of the

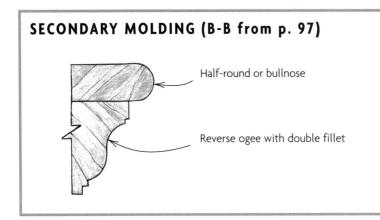
INSTALLING THE MOLDING

When installing a molding, I usually "walk it around," starting at one end of the mantel and butting one miter next to another until I reach the other end.

Begin with a crisp 90° end cut and place it against the foundation board. Then mark the molding for the first miter cut. After making the cut at the mark, check it for a good fit against the pilaster. Don't attach it until the second, or corresponding, miter has been cut and checked for fit. Carefully work your way around the pilaster and across the mantel, taking care at each miter to achieve a tight fit and a perfect match (see the photo on the facing page). This type of interior architectural work attracts a lot of close inspection, so don't compromise the project by settling for sloppy miters crammed with wood filler.

The large cove molding is the first installed, and it goes directly beneath the shelf. Next come the upper bead, the checkered band, the lower bead, and the small cove at the bottom.

entablature (a collective term for all the moldings and treatments that run horizontally along the top of the mantel and are supported by the pilasters). Second, it serves as a capital, or head, for the pilasters.

The ½-in. bead projects less than the other beads, but it doesn't have anything above it. The surface of the beaded molding is left bare, creating a shelf.

Quirk-and-bead molding trims out the inside of the foundation (see the drawing on p. 97). This is a popular and attractive treatment that highlights and softens the edge. It also serves as a filler piece against walls and brickwork to conceal minor gaps.

BLOCKING OUT THE CORNICE

Sometimes in order to achieve proper clearance for a molding and for a more attractive vertical progression, the large cove of the cornice must be blocked out. To achieve this and to give the cornice proper visibility and prominence, the cove is installed over a foundation or platform that will not be seen. For blocking, you can use strips of scrap wood, glued and nailed into place. If any portion of the blocking will show, use clear wood and miter the outside corners. Inside corners can simply be butted together. On this mantel, the central portion of the cornice is blocked out just over the central panel with a panel of solid pine (see the drawing on p. 103).

Each successive molding is snugged up to its neighbor above. This part of the project will take a little time. Don't hurry; do what you need to do to get it right.

Cutting the miters

The miters can be cut either on the table saw using a miter gauge, with a power miter saw, or by hand. The power miter saw offers speed, ease, and accuracy, so that's what I use. The only difficulty with this is cutting the smallest pieces for the inside miters. The rotation and speed of the sawblade can catch hold of the small pieces, trapping them between the rotating blade and the fence and causing them to ricochet across the shop. For these small pieces I recommend using a small miter box and a handsaw.

Start with the large cove under the shelf. Cove molding can't be cut on the flat because when it is installed, pitched at a 45° angle, the ends will not meet up properly. It should be set up and held in position as it will sit on the mantel before being cut at 45°. Use a jig (see the drawing at left) to support the cove molding in the correct position when making your cuts.

"Walk" the moldings across the mantel and around the pilasters, fitting, measuring, and cutting as you go. The smallest pieces can be glued and held in place with masking tape, instead of nails.

MITER JIG FOR COVE MOLDING

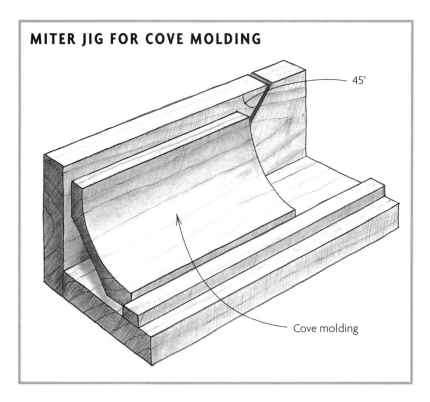

45°

Cove molding

Tips for nailing

In the 18th century, small, thin cut nails were used to attach the moldings. These finish nails had a distinctive rectangular head, and because of their flat tapered shank they were less likely to split the molding than round (wire) nails. If you decide to attach the molding with wire finishing nails, be sure to predrill the holes in order to avoid splitting the molding. Also use a nail set to sink the nails below the surface of the molding.

For much of my finish work, I used two types of air-powered nail guns—a brad nailer for the thicker moldings and a pin nailer for the smallest moldings and for pinning the miters. Nail guns are an excellent method for attaching finished moldings. They are quick, requiring only the pull of a trigger to drive a nail completely beneath the surface. They generally don't split the wood, and because a hammer or nail set isn't used, there are no hammer dings or large holes to repair later. Another benefit is that you can position the workpiece with one hand while using the nail gun with the other.

Short lengths and small thin moldings often split when nailed. The best way to avoid this problem is to glue them in place and not nail them at all. Masking tape can be used to hold them until the glue dries.

To minimize sanding and filling of any exposed nail heads, try to fire a nail into the quirk or recess of the molding, if you can reach it easily. In some situations, pinning the molding through the crown of the bead gives a better hold and causes less damage to the molding.

Place your nails about 3 in. apart and stagger them to help hide the heads and to improve holding power. After wiping any excess glue with a damp rag, apply a wood filler to cover over the nail holes. Then sand everything flush.

BUILDING THE SHELF

With the decoration applied, all that's left is to add the shelf. For the shelf on this mantel, use 5/4 pine. The edge of the shelf echoes the outline of the moldings beneath it (see the drawing on the facing page). It bumps out over the pilasters and the central panel. Instead of starting with a wide board and cutting away material to form the outline, you can glue narrow strips of pine onto the front edge of a narrow board to create the projections. Then rabbet the front edges to slim them down visually. The back edge of the shelf should also be rabbeted to make scribing the shelf easier later. Make sure the rabbet doesn't run straight through, or it will be visible from the ends when the shelf is installed.

The shelf's edge can't be molded with a plane because of the broken front edge and the end grain. In the 18th century, the customary method was to use a plane run off the molding on a different board, cut it free, and attach it to the mantel shelf with small cut nails and hot hide glue. You can do essentially the same things using modern power tools: a router to run off the molding, a table saw to cut it free from the board, and an air-powered finish nail gun and yellow glue to attach it to the shelf's edge. You can't just rout the edges of the shelf directly because the rounded inside corners would be a dead giveaway that the shelf was made by modern means.

The shelf profile on this mantel is made up of a $\frac{3}{8}$-in.-dia. half-round with a $\frac{3}{8}$-in. cove beneath it. This combination provides a soft edge that leads into a cove. This molding, which is applied to the shelf's edge, breaks up an otherwise thick square edge and at the same time reinforces the shelf's horizontal line.

If you decide to make a bow-front or curved mantel shelf, the molded treatment on the edge of

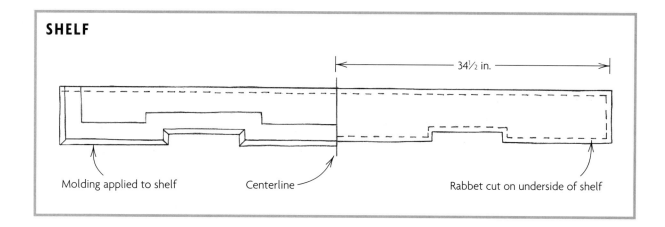

SHELF

34½ in.

Molding applied to shelf   Centerline   Rabbet cut on underside of shelf

the shelf should be routed, since its shape is an outside curve, completely accessible to the router bit. But if a true period treatment is desired, use the plane method and place the strip in a shower under hot water for about 15 minutes. This will soften it and make it pliable. Tape or tack the molding onto the edge of the curved shelf, without glue, until it dries. This will accustom the molding to the curved edge, making the final gluing and nailing much easier.

INSTALLING THE MANTEL

The mantel shown in the photo on p. 96 is installed over a brick-fronted fireplace and attached directly to the brick. Since the mantel itself wasn't heavy and no load would be placed upon it, I secured it to the brick with hardened masonry screws. These fas-

teners, which look like blue drywall screws, require a small pilot hole drilled into the masonry before being driven into the brick with a heavy-duty screw gun. The four ½-in.-dia. holes drilled into the mantel to a depth of ⅜ in. were plugged later to hide the screws.

If the mantel is being installed over an old fireplace with a wooden lintel and wooden anchor blocks, drive small-head trim screws through the mantel and into the lintel and anchor blocks. The heads can be covered with putty later. If you have to install any heavy-duty fasteners (like lag bolts and lead anchors), plan to locate them under a strip of molding if possible.

The installation of this project mantel in the New York City apartment is typical of a modern installation. The mantel is attached directly to the masonry with countersunk screws. If in

your own house a wood finish and modern design prevents this direct approach, you can screw wood cleats to the masonry and mount the mantel over them, making the attachment with glue and/or small, unobtrusive fasteners.

SCRIBING AND INSTALLING THE SHELF

Before you can install the shelf, you'll need to scribe it to the wall. (Any small gaps between the foundation and the wall can be eliminated with caulk.) Once the mantel is installed, start by setting the shelf on top of the mantel and against the wall, centering the shelf and positioning it for a uniform overhang all around. Now measure the widest gap along the back between the wall and the edge of the mantel, and set a compass for the width of the gap. Place the pin of the

SHOOTING MITERS FOR A PERFECT 45°

Over the years, I've examined hundreds of mitered joints on early furniture and architectural woodwork, and I've been impressed by their precision and tight fit. To me, this was proof that early woodworkers weren't joining their miters straight off a miter box and backsaw or dovetail saw. So how were they achieving such clean and accurate joinery?

In the 18th and 19th centuries joiners used several "appliances" to dress rough-cut miters. Some, like the miter jack, were purchased from tool suppliers. Others, like the shooting board and donkey ear, were made by the craftsmen themselves. These simple devices are effective today as they were centuries ago.

After cutting the material on the miter box, joiners would use a miter jack to dress large moldings. This fixture (see the photo below) firmly supports the rough-cut workpiece between two adjustable 45° ramps; it's basically a vise with ramped faces set onto a frame. A low-angle miter plane was then set upon the ramps for support and passed over the rough-cut miter. A few fine passes left the face of the miter as smooth as glass and at a perfect 45°.

The shooting board, another device for trimming small miters, employs a fence with a 45° end attached to a stepped-down bed (see the top photo on the facing page). Typically shooting boards were used for moldings that have their miter across the width of the board, with the edge cut at 90°, such as for a picture frame. The mitered molding is held against the fence and the plane is pushed on its side, along the bed and across the mitered end. Small moldings don't have to be clamped; they can be kept in place by hand pressure alone.

For the smallest moldings, joiners could use a jig called a donkey ear. A donkey ear is useful for dressing edge miters, such as those wrapping around corners. It has a fence, which supports the molding material, angled up away from the bed at 45°, allowing the plane, resting on its side, to pass over the miter cut. The orientation of the molding to the bed brings the plane's blade into immediate contact with the full surface of the miter (see the bottom photo on the facing page).

A miter jack clamps a piece of molding between two ramps at the proper angle to be planed to a perfect 45° cut.

compass to follow the contour of the wall, while the pencil end of the compass transfers that contour onto the top of the shelf.

Use a sharp spokeshave and block plane to shape the edge of the shelf to the pencil line. When you're done, check the fit. After a few adjustments, the shelf should rest tightly against the wall. Drill small pilot holes in the shelf every 18 in. for Phillips-head trim screws. Using a ⅜-in. heavy-duty electric drill, drive four screws into the top edge of the foundation and the tops of the pilasters, just below the surface of the shelf. Later fill the holes, sand the surface flush, and apply the finish of your choice.

FINISHING TOUCHES

This particular mantel was stained a very light color, sealed with a shellac wash, then brushed with a pickling mixture (white Japan oil pigment mixed with turpentine). The pickling mixture was used to create the impression that the mantel was originally painted, then stripped to the wood. It leaves the wood with an attractive overall pale color that allows the wood grain to show through. Later, the mantel was coated with paste wax and polished.

With the aid of a shooting board, you can use a miter plane on its side to shoot a crisp miter on small moldings.

The donkey ear is a simple but handy device that can be quickly cobbled together from scrap in a few minutes. With it you can plane miters on very small pieces of molding.

DOOR AND WINDOW CASING

Moldings are important because they create order and establish scale within a room. They also give a room style and reveal its importance within the house. In the 18th and 19th century, the purpose of window and door casing was to provide a stop for the plaster or paneling around the windows and door openings and to dress the joint between the jamb and the wall. Sometimes a flat, plain board functioned as trim; other times, more elaborate moldings were crafted.

Early door casing and trim was robust and thick. If you took a section of an early door casing and analyzed its design, the first thing you would notice is its pleasing mass. Then you would notice how it rises in thickness from the jamb (inside) edge in a series of platforms punctuated by beads and ogees. Finally, you might hold the molding near a window and see how the light plays over its contours—the relationship of light to shadow and flat to curve can transform a length of molding into sculpture.

Plain casings, too, were wide and thick. Since molding was produced by hand, its dimensions would be suited to the height, width, and design of the door. So even a plain treatment would appear substantial.

Nowadays, moldings are given short shrift. I remember, as an apprentice, rushing through a trim job in assembly-line fashion: one carpenter laying out, another making the cuts, a third nailing the trim, and a fourth applying caulk and filling nail holes. It was brutal, boring work and it showed in the finished job. Modern mass-produced moldings are full of cuts and details, but virtually all are in the same plane (see the photo below). There's no real relief to the surface, no significant change in elevation to give emphasis to the details. Unfortunately, this is what passes for a Colonial-style or Georgian molding today.

This chapter's project, a window casing that can be adapted for use on doors, was inspired by the door and window casing in my 1812 farmhouse. The original casing is a single, stepped-down length of molding that was cut to overlap in a way that virtually eliminated any cross-grain movement and subsequent shrinkage (see the sidebar on pp. 120-121). That sort of work calls for a master's skill, but there's a faster and easier way to achieve the same look, and that's by building up the molding using separate pieces.

The "Colonial" molding commonly used in traditional homes nowadays is a thin and shallow imitation of the real thing.

During the 18th and 19th centuries, interior woodwork was made almost exclusively of pine, and for good reason. Pine was easy to harvest, mill into boards, and prepare as molding. Back then, it was important for the wood to be relatively clear and free of knots and blemishes in order to produce the best finished molding. With today's high-speed shaper cutters and router bits, however, material with small knots can be safely used. Your only concern should be about any bleed-through on the finished paint job.

THE PROJECT'S PARTS

This project has four components (see the drawing below): the primary foundation molding, the secondary foundation molding, the bed molding, and the backband.

BUILT-UP WINDOW AND DOOR CASING

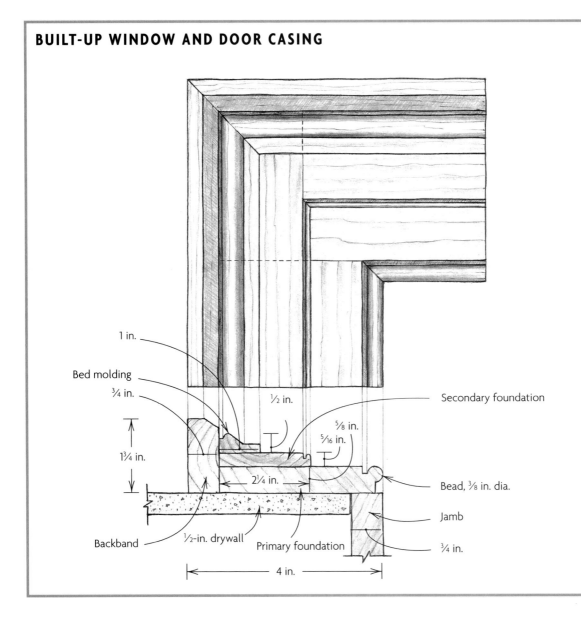

1 in.

Bed molding

¾ in.

½ in.

Secondary foundation

1¼ in.

⅝ in.

5/16 in.

Bead, ⅜ in. dia.

2¼ in.

Jamb

Backband

½-in. drywall

Primary foundation

¾ in.

4 in.

The primary foundation molding is flat and has a full bead with a return along the inside edge. When properly installed, the bead looks like a dowel set into a corner and gives any prominent outside corner an unusual touch. The bead decorates the junction of the jamb and the trim, and it also hides any irregularities at this joint. On top of the primary foundation molding is the secondary foundation molding. This piece also is flat, but has a small bead along the inside edge. The secondary foundation molding breaks up the wide expanse of the primary foundation molding with a change in elevation. The next layer, the bed molding, rests against the backband, which projects off the foundation molding and defines the perimeter of the casing. The backband's profile is softened by a 45° bevel on the inside edge, but it could be left square.

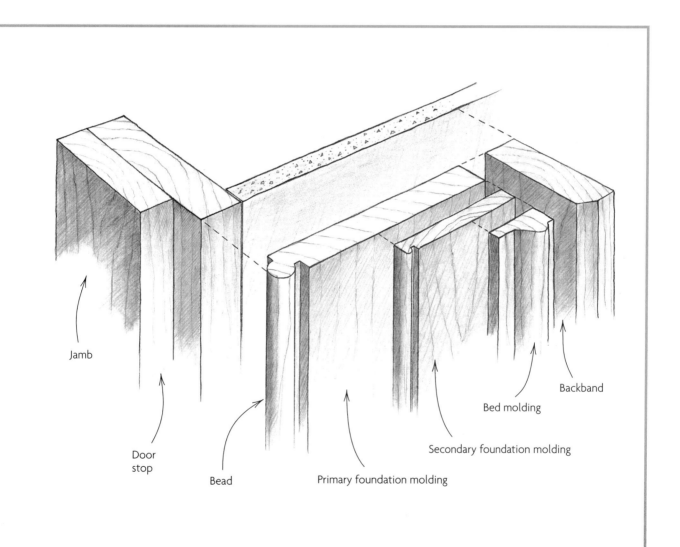

Jamb

Door
stop

Bead

Primary foundation molding

Secondary foundation molding

Bed molding

Backband

The fastest, safest, and most uniform way to mill molding is on a router table equipped with a dust-collection port.

MILLING

I mill my moldings on a router table equipped with a 1½-hp router. The stock can be passed quickly along the fence and over the exposed bit, increasing my feet-per-minute rate and improving the accuracy and uniformity of the molding profile. The router table I use has an effective dust-collection system integrated into the fence, which catches about 95% of the chips and dust (see the photo above).

I won't describe in detail how to mill the backband, the bed molding, or the secondary foundation; all of them are simple to run off in the shop using readily available bits. You might even be able to buy them at a lumberyard. The primary foundation molding, however, is a little unusual. The pattern is produced in two simple steps on a router table using a ½-in. beading bit (see the drawing on the facing page). The first cut is made on the face of the board. First, thickness the stock to ⅝ in. Then, with a trial molding piece laid on edge, set the beading bit

flush with the router table. This will produce a ½-in. bead with a ⅛-in. quirk (recess) on the face side. Check the sample, then run off the rest of the material.

For the second cut, place the trial molding piece face up and visually align the beading bit with the previous cut. Run off your sample. The finished trial piece should look like a ½-in. round bead with quirks on each side. Make sure the completed bead is fully formed without any flat or misformed areas. If the trial piece looks good, run off the rest of the material.

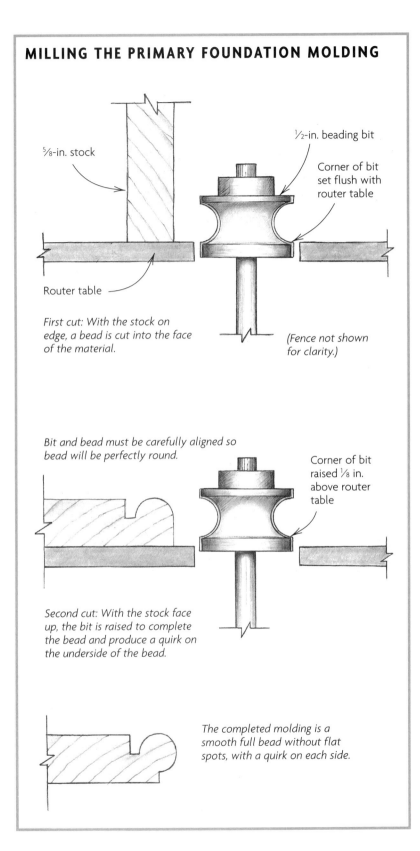

MILLING THE PRIMARY FOUNDATION MOLDING

⅝-in. stock

½-in. beading bit

Corner of bit set flush with router table

Router table

First cut: With the stock on edge, a bead is cut into the face of the material.

(Fence not shown for clarity.)

Bit and bead must be carefully aligned so bead will be perfectly round.

Corner of bit raised ⅛ in. above router table

Second cut: With the stock face up, the bit is raised to complete the bead and produce a quirk on the underside of the bead.

The completed molding is a smooth full bead without flat spots, with a quirk on each side.

A MITER FOR WIDE MOLDINGS

The most common problem with wide casing is that the miter tends to open up as the wood dries. No matter what pre-cautions are taken, this condition is almost unavoidable. I've gone so far as to use veneered plywood edged with solid wood to ensure a tight miter. The project treatment in this chapter avoids the problem and at the same time produces an attractive variation of a period joint. How does this joint keep from opening up? Unlike a full 45° degree miter, the miter on this right-angle joint is limited to the width of the ½-in. bead and quirk. The rest of the material is cut at 90°.

Begin by marking out a 45° cut, 3 in. (the width of the molding minus the bead and quirk) from the end of the horizontal pri-mary foundation. On the band-saw, cut along the quirk until you reach the marked-out miter (see the top photo on p. 116). Using a mitered guide block and a dovetail saw, cut off the bead at 45° (see the bottom photo on p. 116) and dress the cut with a chisel. The other half of the joint, the vertical primary foundation, is cut to the proper length at 90°, then the bead is mitered. Joined together, the two pieces look good: a clean, precise

On the horizontal piece, the bead and quirk are trimmed off, then mitered. The miter should begin at a measurement equal to the width of the molding, minus the quirk and bead.

The easiest way to miter the bead and quirk is to use a dovetail saw and guide block. The cut can be cleaned up with a wide chisel.

butt joint embellished with a perfectly mitered bead (see the photo at top left on the facing page).

For the secondary foundation, repeat the cuts used on the primary foundation molding, but stagger the pieces so they overlap (see the photo at top right on the facing page).

INSTALLING THE CASING

The primary foundation is attached first to the window or door jamb and extends up to the existing plaster or over the drywall (see the drawing on pp. 112-113). Placement is critical. The quirk on the underside of the bead should be set back from the jamb so that the bead is flush with the jamb, creating a small reveal. The effect will be of a full-round bead placed into a square recess. Later the outer edge is nailed through the drywall and into a stud, with nails spaced every 24 in.

Make sure the bottom end of the casing is square to the sill by inserting shims between the molding and the wall (see the bottom photo on the facing page). Now you can toenail or screw the primary foundation molding in place. Be sure to place the fastener where it will

Careful cutting produces an attractive miter and butt joint.

The secondary foundation is cut just like the primary foundation, except that the parts are staggered to overlap.

Here, a shingle is used to shim the window molding in proper position, so a screw can be driven at an angle into the chair-rail sill. Later, the screw will be covered by the backband.

The bed molding is mitered and laid against the backband and the secondary foundation molding to complete the pattern.

be covered over later by either the backband or the secondary foundation molding.

If you are nailing manually (using a hammer), insert and drive the nails into the quirk recess on the face of the molding and countersink them. If you are using a pneumatic nail gun, shoot through the center of the bead, at a slight angle, into the jamb. The nails should be spaced about 6 in. to 9 in. apart along the molding's length and 1 in. from each end.

Don't glue the secondary foundation molding onto the primary foundation molding. It is suffi-cient to secure it with a minimal number of pneumatic or wiring finishing nails. Later, these nail holes can be puttied over.

Next, miter the backband and nail it onto the edge of the foundation molding.

The last molding to install is the bed molding. After mitering, place it snugly against the backband and the secondary molding, closing any gaps (see the photo above). Each of the successive moldings is placed to overlap and secure the previously attached layer. So try to place your fasteners where they'll be covered by the next layer of molding in order to avoid filling nail holes later. The resulting assembly will be beautiful, strong, and forever gap free.

A WINDOW APRON

An apron should be installed underneath the window sill; without one, a window seems unfinished and unbalanced. Usually the apron is a simple horizontal member, set under the sill to give some visual weight to the window trim. Often, it's nothing more than a piece of the window-casing molding. But, sometimes it's an unusual and dramatic touch that grabs the viewer s attention. I like to have fun with this part of the window trim, so I try to come up with something eye-catching and a little different.

In my own house, the challenge was to design an apron that would be attached under the chair rail and end in line with the window casing. The apron had to be substantial enough to hold its own against an elaborate chair rail, yet not overpower it visually. I decided on a simple apron with two moldings: a cove along the bottom edge and a raised bead just above it. These simple moldings cast a strong shadow in a horizontal line and establish a nice counterpoint with the staccato line and independent elements of the chair rail.

For the apron, I used a ⅞-in.-thick piece of clear pine (the rest of the casing is built from ⅝-in. stock). It had to be thick enough to project off the wall sufficiently and provide a little room for a generous return. For balance and to establish the apron as the base of the window, I used a piece of 2½-in.-wide wood. Along the bottom edge of the apron, I attached a ¾-in. cove molding, then coped and profiled the ends. About ¼ in. above the cove, on the face of the apron, I laid a raised bead. I mitered the molding at the corners so the profile would die into the wall. The apron assembly is shown in the photo and drawing below.

The window apron looks like a complicated molding, but is really only a raised bead and a cove, as the drawing at right makes clear.

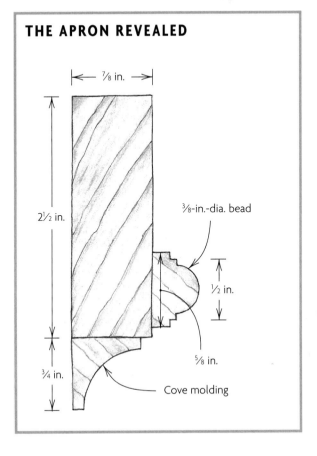

THE APRON REVEALED

⅞ in.

2½ in.

¾ in.

⅜-in.-dia. bead

½ in.

⅝ in.

Cove molding

STEPPED MITER JOINT

My house is more than 185 years old, and when I moved in it showed its age: tilting floors, sagging ceilings, cracked plaster, and loose floor boards. However, every single joint on the window and door casings was perfectly fitted and tight. Because they were covered with layers of thick paint, I really couldn't see how the casings were put together until I took one apart. I had never seen a joint quite like it before.

The joint wasn't a straight miter. It was more like a combination of small miters and straight butt joints. The basis of the joint is the stepped foundation molding, which was made of a single piece of wood. On one half of the joint, the upper step was removed (see the photo below), leaving it flush with the lower step. The large bead was cut away and mitered. On the other half of the joint, the bottom half of the foundation molding was cut away, leaving the upper step projecting onto the first piece (see the top photo on the facing page).

To ensure perfect registration of the perpendicular surfaces, the mating inside ends were cut and pared to butt at an angle (see the bottom photo on the facing page).

To remove the upper step, a series of shallow saw cuts is made on one half of the joint, so the waste can be pared away. A segment of the bead is removed and mitered.

On the other half of the joint, the cuts are reversed. The bottom portion of the molding is cut away, leaving the upper part to overlap the reciprocal half.

Angling the inside ends of the joint ensures a perfect fit.

10

FRAME-AND-PANEL
DOOR

A front door does more than keep out the flies in summer and the cold in winter. It is the single element in the doorway that draws everything else together into a cohesive design. It is the key to the rhythm of the doorway—and possibly to the house.

One of the problems with a standard manufactured door is that it lacks rhythm. By that I mean there is no flow to its design. A door doesn't need yards of carved molding or other detail to be captivating, but its basic design and craftsmanship should engage the viewer's eye. When I look at a door, my eyes move over it searching for a pattern, trying to establish a relationship between the different parts. This is what makes a door interesting and sets it apart from others. A door laid out with uniform panels and rails is boring. There's nothing for the mind and the eye to do but count panels.

Consider how doors are manufactured today. In a factory, it's economical to glue up wide panels from narrow boards, then cut them to the same size and mold them to the same pattern on both sides. Rails ripped to the same dimensions can be interchanged easily, not only on the same door, but on other designs too. The manufacturing process is simplified, and the

cost is reduced. This might produce a door that is acceptable, but not exciting.

A door that you design yourself, following the guidelines presented here, can make all the difference in the exterior appearance of your home.

The door in the drawing on p. 124 is typical of early 19th-century doors. There are two important features in its construction. The first is that wide expanses of wood are avoided. Wood exposed to seasonal extremes of temperature and humidity will change dimension across the grain. In warm, humid weather, wood swells. In cool, dry weather, it shrinks. If free movement is prevented, the door will eventually warp or buckle and break apart. By constructing the door as a frame and filling the frame with an arrangement of small panels, the wood can move without constraint.

The second feature is the through wedged mortise-and-tenon joint. A through mortise-and-tenon joint has a portion of the horizontal member, which is called the rail, penetrating the vertical member, called the stile. Mechanically, even without the benefit of glue, this joint will stay together. When every major joint on a door is a through

joint, you have a door that will remain tight, square, and flat for hundreds of years. In contrast, many manufactured doors are assembled with dowels or short tenons. Given the great weight of a door and the width of the threshold it must span, dowels will not hold a door together or square for long.

A frame-and-panel door has three main components: stiles, rails, and panels. In frame-and-panel construction, a narrow groove milled along along the inner edges of the stiles and rails receives the panels, which fit snugly but are left unglued. That way, they are free to move in response to their environment without causing damage to the door.

The stiles, or outer vertical members, determine the height of the door and receive most of the mortises. Besides the two outer stiles, which run the length of the door, there may also be a number of shorter interior stiles, or muntins. The width of the stiles will depend on the design and size of the door, but it is generally between $3\frac{1}{2}$ in. and 8 in. The thickness of the stiles also varies greatly. I have several delicate paneled doors in my home that are only $1\frac{1}{8}$ in. thick, but my own entry door measures a full $1\frac{7}{8}$ in. The door in this project is $1\frac{1}{2}$ in. thick.

THE FRAME-AND-PANEL DOOR

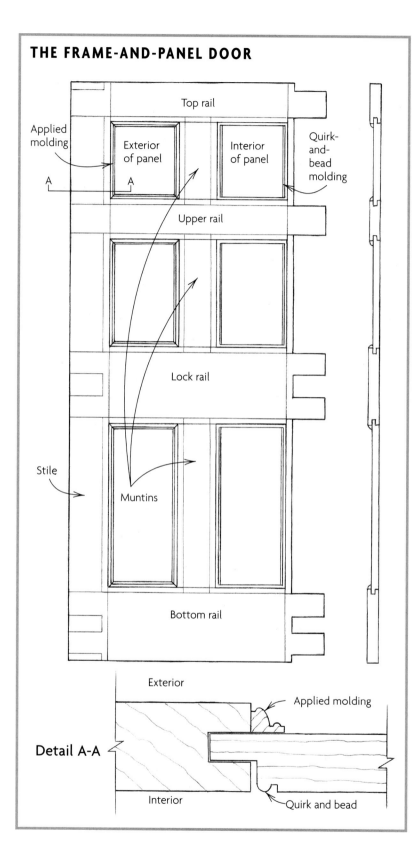

Top rail

Applied molding

Exterior of panel

Interior of panel

Quirk-and-bead molding

A

A

Upper rail

Lock rail

Stile

Muntins

Bottom rail

Detail A-A

Exterior

Applied molding

Interior

Quirk and bead

The rails are the horizontal members that fit between the stiles. Most doors have top, upper, lock, and bottom rails. These rails usually vary in width. Generally, the width of the top and upper rails is equal to that of the stiles. The lock rail is usually the widest, for two reasons. As the central supporting member, it prevents the door from sagging. In addition, many old homes had a large carpenter's lock or brass box lock that required a large surface for installation, hence the name. The bottom rail is often a little narrower than the lock rail, but still wide enough to keep rain and snow out of the panel grooves.

The panels are sized to fit the spaces left between the stiles and rails. Some panels are only as thick as the width of the groove in the frame members, and their surface is left plain and flat (see the drawing on the facing page). Others are raised or fielded on one or both sides (see the sidebar on p. 133-135) and can be as thick as the frame members. On early doors, panels were usually made from a single piece of wood, but today panels are often glued up from narrow strips. This method is said to produce superior stability and control wood movement. However, in a paneled door, that's the function of the frame. I think it's simply an unattractive cost-cutting measure.

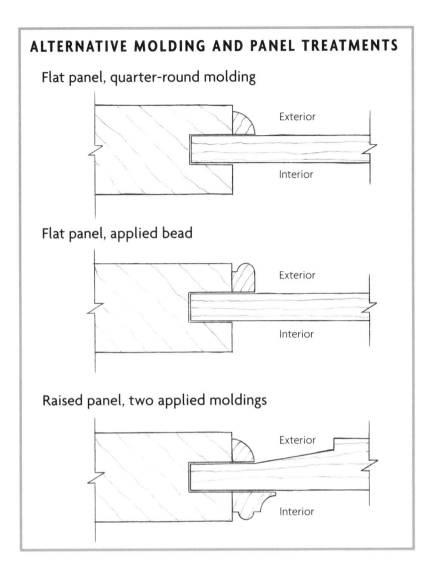

ALTERNATIVE MOLDING AND PANEL TREATMENTS

Flat panel, quarter-round molding

Exterior

Interior

Flat panel, applied bead

Exterior

Interior

Raised panel, two applied moldings

Exterior

Interior

DESIGN

The exact arrangement and size of the panels in this door were in part determined by the dimensions of the frame stock. The widest 1½-in.-thick pine I could obtain was 12 in. After jointing and cleaning up the edges, it yielded material 11¼ in. wide, which was adequate for the lock and bottom rails and could be ripped in half to form the stiles and muntins with a minimum of waste.

The width of the lock and bottom rails is critical to the performance of the door. A wide lock rail, with double tenons, helps to hold the door together and provides a generous ground for the door's hardware. The bottom rail, which also has double tenons, raises the panels off the ground, protecting them from rain and snow and also visually anchoring the door.

The width of the panels remains constant at about 10 in. throughout the door, with their length increasing from top to bottom. The top panels are almost square.

Doors from the 18th century were usually fitted with raised panels, but early in the 19th century flat panels became popular. As shown in the detail of the drawing on the facing page, these flat panels were recessed on the exterior side with molding applied at the junction. On the interior side of the door, the panels were often flush with the door frame with a small reveal on all four sides and a bead and quirk along the edges. The bead-and-quirk flush panel is fairly common in period examples and is extremely attractive, but you never find it nowadays in commercially available doors.

MATERIAL

Almost all the early doors in domestic architecture in the 18th and 19th centuries that I have examined were made of pine. Back then pine was plentiful and cheap. Trees yielded wide boards, and pine was easy to mill. Today, Select pine costs more than some hardwoods, but I still choose it over anything else. I choose the best commercial grade and then I go through the bins and pick the best of the best, searching for light weight (indicating low moisture content), straightness of grain, and no twist along the length. For an exterior door, I like to start with material that is 1¾ in. to 2 in. thick.

I have always felt that the extra money you spend on good-quality material will be saved in labor later on. Building a door of clear straight pine also means cleaner cuts and better surfaces when using both hand and power tools. Straight-grained pine also stays that way. It's less likely to warp or twist later.

MILLING

After jointing and thicknessing the material to 1½ in., select the planks with the straightest grain for the stiles and rip them to final width and length. You can cut the muntins from material that is not quite straight enough for the stiles. Cutting these pieces to a shorter length will reduce the effect of twist or other deficiencies. Rip the rails to width, but cut them ¼ in. over final length to allow the ends of

the tenons to project slightly through the mortises, as described on the facing page.

CUTTING THE GROOVES

The grooves that will hold the panels should be ½ in. wide and ¾ in. deep, centered on the edges of the rails, stiles, and muntins. To cut them, set up the table saw with a ½-in. dado head for a ¾-in.-deep cut. Run a test piece through the setup to make sure that the groove is dead center, which will enable you to maintain perfect registration if you have to flip a workpiece in the door for a better-looking appearance. When the test cut is satisfactory, mark the rails and stiles for grooving, and run all the stiles, muntins and rails on edge over the ½-in. dado head.

With all the material cut to length and grooved to receive the panels, the mortises can be laid out.

MORTISE-AND-TENON WORK

Since this door will be 1½ in. thick, the mortises and tenons should be one-third of that thickness, or ½ in.—the same width as

the grooves. This works out conveniently, since the mortises start inside the grooves. A thickness of ½ in. also leaves plenty of material surrounding the tenon, and the tenon itself, because of its width, will be substantial.

A proper mortise is closed at each end to hold the tenon fast and square. On a door I leave 1 in. at the ends of the stiles to contain the tenon and keep it from slipping at glue-up. However, if the tenon was only as wide as the mortise was long, a good portion of the rail's width would be unsupported and liable to twist or open up at the joint. Haunching the tenons (see the top photo on the facing page) supports the top and bottom rails and keeps them from twisting. This is also is a convenient way to fill in the exposed ends of the groove.

Since the stile and rail parts are the same thickness and expected to join up flush, all the tenons are cut to the same thickness. However, each of the rails performs a different function, and so each tenon is configured differently. The design takes into account the function, the width of each rail, and the subsequent wood movement, which, if not accounted for, could cause the door to sag, split, or break apart.

The haunch on the top and bottom rail tenons helps keep the door square, strengthens the mortise, fills the end of the groove, and keeps the shouldered portion of the tenon flush with the stile.

To cut the tenon on the table saw, raise the blade to just under the layout line for the groove. Later, the tenon will be trimmed to final thickness.

The top rail has a haunched tenon. The through-tenon portion of this rail measures the width of the rail, minus the haunch (in this case, 1 in.) and the depth of the groove. The critical dimension is the space between the mortise and the end of the stile, which is referred to as the set-in. For an entry door, a 1-in. set-in will keep the sides of the mortise from blowing out and provide the structure necessary to keep the joint square.

The upper rail has a simple tenon; its width is the width of the rail minus both grooves. No special measures need be taken.

The lock rail has a double, or crenelated, tenon. Because of the great width and central location of the lock rail, its tenon must be carefully dimensioned. Unlike a single wide tenon, a double tenon permits wood movement within the joint, and mechanically two tenons are stronger than one. The shared set-in helps keep the joint rigid and square.

The bottom rail has a double haunched tenon. The double tenon is needed because the rail is very wide, and the haunch strengthens the joint and also fills in the exposed groove at the end of the door stile.

Cutting the tenons

After changing from the dado blade to a regular combination blade, mark out the thickness of the tenon (½ in., in this case) on the ungrooved edges and the ends of the rails. Next mark out the length of the tenon plus ⅛ in. around the four sides of the rail, using a square and a sharp pencil or marking knife. Then raise the blade of the table saw to just under ½ in., using the line scribed onto the end of the rail as a guide (see the bottom photo at left).

Kerfing is a quick and safe way to ensure uniform thickness of the tenon cheeks. The waste between the kerfs can be removed with a chisel, and the cheeks can be dressed with a shoulder plane.

Tenons can also be cut on the bandsaw, as long as their width does not exceed the resawing capacity of the saw.

With the rail supported by a miter guide, make repeat cuts across the tenon (see the top photo at left). It's easier to achieve uniform depth of cut and therefore uniform thickness of the tenon by making repeat cuts instead of using a dado. Pay particular attention to the shoulder cut: If it is cut out of square, the door will be difficult, if not impossible, to square up. To trim the waste pieces off the tenon, use a large chisel or a shoulder plane.

Another method for cutting the tenon is to cut the shoulders on the table saw, then cut the cheeks on the bandsaw (see the bottom photo at left). However, with this method you can't cut tenons that are wider than the resawing capacity of your bandsaw.

Whichever method you use, cut the tenons a little fat so there is a way to compensate for any drift or error when cutting the cheeks. The tenons will be trimmed to a nice fit later, as described on pp. 129-130.

For the lock and bottom rails, cut two narrow tenons with a haunch running between them (see the drawing on the facing page). These rails are wider than the top two, and this configuration allows the rail's width to move without causing a split or crack to develop. In a very wide

LOCK-RAIL TENON

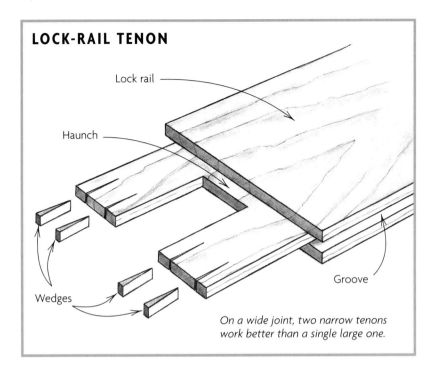

Lock rail

Haunch

Wedges

Groove

On a wide joint, two narrow tenons work better than a single large one.

mortise-and-tenon, the glue joint will eventually break apart because of the movement taking place within the joint. Mechanically, the joint won't stay together because the thin walls of the wide mortise will flex and move apart from the tenon.

Like the upper rail and the lock rail, the muntins are grooved on both edges, then a ¾-in. stub tenon is cut on each end. Since the muntins are securely contained within the frame, a short tenon that fits into the groove of the rail is sufficient.

After cutting the muntins to length (the distance between rails plus the depth of the grooves), set the sawblade con-

servatively for a ½-in. depth of cut and the fence for a ¾-in. shoulder and make the four shoulder cuts. Next, raise the sawblade to ¾ in. and set the fence for the cheek cuts. The pieces are set on end and carefully passed over the rotating blade. Cut the cheeks a little fat, then trim and fit them later.

Cutting the mortises

Lay the rails in place along the door stiles and mark out the exact length of the mortise off the tenon, then transfer these marks around the sides and edges of the stile.

Mortises must be straight and clean without tearout, so rather than trying to drill your holes

through in one shot, drill in partway from each end of the mortise instead. Any minor misalignment of the holes inside the mortise can be cleaned up later with a planemaker's float or wide chisel.

The mortises can be cut with a hollow-chisel mortiser, providing the capacity of the machine will accept the width of the stiles. A clean job can be also achieved on the drill press using a ½-in. brad-point bit.

The top and upper rails get one long mortise; the lock and bottom rails get two narrow mortises, with the stub tenon between them.

Fitting the joint

Since the tenons are cut a little fat, you will need to trim them to fit the mortises. For this work I use a heavy English shoulder plane, but a modern Stanley or Record shoulder plane would work just fine. If you can't obtain a shoulder plane, a Surform tool will work too.

I trim my tenons so they're full at the end to ensure a gap-free fit where they penetrate the stile, but a little light in the middle to allow easy engagement, and then a tad tight at the shoulder to bring everything in nice and flush. The only tool that can give me that kind of control is my English shoulder plane. This plane has a blade that spans the full width of the plane, allowing it to cut into corners. The small mouth delivers a fine, smooth surface, even when planing across the grain. And the weight completely eliminates any chatter. With a tool like this I can trim a tenon to perfection in just a couple of minutes.

After trimming the tenons, you can kerf them to receive the wedges (see the drawing on p. 129). With a backsaw, cut two kerfs about ½ in. in from the edge and ¾ in. deep, and set the rails aside while you work on the panels. The kerfs in the end of the tenons will be wedged at glue-up.

PREPARING THE PANELS

Thickness the panel material down to 1 in. Select wide straight-grained panels, and avoid any unnecessary seams and joints. If panels must be glued up, try to camouflage the seams by placing them at straight-grained portions of the panel. Indiscriminate gluing of narrow boards to produce wide panels will produce distracting surface patterns that eventually telegraph through a paint finish.

After dry-fitting the stiles and rails and checking the frame for square, measure the actual space

between the frame members for your panels. To those figures add the depth of the grooves (¾ in. in this case). The total will be the untrimmed size of the panel. To allow for cross-grain expansion of the panel in warm, humid weather, take about ¹⁄₁₆ in. off each edge with a hand plane.

On all four sides of the panel, cut a tongue that's ½ in. thick and ⅞ in. long. This tongue will fill the width of the groove and produce a ⅛-in. reveal (the quirk) between the panel and

the frame. This quirk outlines the panels on the interior of the door and helps to hide any seasonal shrinkage that occurs.

When the panels are prepared, dry-fit the door frame again, but this time with the panels in place. Using a shoulder plane, adjust and trim any panels that prevent the frame from fitting together. Check the quirk around the panels for uniformity. Once the frame and panels fit properly, check the door for squareness and wind (twisting). Also make

The completed panel shows a ¾-in.-wide rabbet with a quirk-and-bead molding cut along the vertical edge.

sure all the frame joints are tight and the ends of the tenons protrude slightly.

Beading

With a ¼-in. beading bit in the router and the router in the router table, set each panel on edge. Raise the beading bit to clear the tongue and cut the profile into the full thickness of the panel. Cut the bead only on the long-grain edges of the panels, which will be vertical in the door (see the photo on the facing page).

You could also perform this operation with a beading plane. Using a plane will give a distinct hand-made quality to the panels. If examined along their length, the beads will meander slightly and the plane might cause small bits of tearout. To enhance the hand-made look, you can also plane the surface of the panels.

A generous number of clamps applied with light, but firm pressure will draw the joints closed without racking the door out of square.

GLUING UP THE DOOR

Now comes the moment of truth. Lay the left-hand stile on edge and coat the mortises generously with yellow glue (use a weatherproof glue for exterior doors). Then insert the tenon of the bottom rail into its mortise in the outer stile. Next, slip a bottom panel in, followed by the lower muntin. (Since the muntins have a short ¾-in. tenon they can be slid into place along with the panels. Make sure that their precise locations are marked on the upper rail and lock rail.) Now, slide the lock rail into place, and then a middle panel. Assemble the door from left to right and from bottom to top.

When the door is fully assembled, check the joints for fit and the door for squareness. Next, check the bead-and-quirk panels for uniform spacing. If they must be adjusted, use a small block of wood and a mallet to tap them lightly into position.

After the door checks out, clamp lightly on both sides of each rail-to-stile joint (see the photo above). Now, before the glue sets and while the joint is clamped up, the through tenons can be wedged. Cut slender wedges from ½-in. stock, coat them with

The through tenons are wedged during assembly (clamps have been removed for clarity). After the glue dries, the wedges will be trimmed and the joint dressed with a low-angle block plane.

glue and drive them gently into the tenons (see the photo above). After the joint dries, saw off the excess and plane the joint flush. If you choose to, you can remove the clamps as soon as the joints are wedged.

FRAMING THE PANELS WITH MOLDING

If your door will have an applied molding on the exterior side of the panels, now's the time to put it in place. Choose or design a molding that will fit against the panel and not project past the surface of the frame (see the drawing on p. 125 for some possibilities). If the molding is recessed, the frame members can be dressed with a plane or sanded after the molding has been applied.

Cut all the pieces of molding on the table saw or with a chopsaw and set them into place around each panel. Cut them snug and dress them with a block plane to ensure tight, gap-free miters.

When all the pieces are in place, apply a small bead of glue to the edge of the molding that rests against the frame. Do not glue the molding to the panel, as this might interfere with the seasonal movement of the panel. When all the molding has been glued, use an air-powered finish nailer to tack the pieces to the frame.

MAKING A REEDED AND NOTCHED DOOR PANEL WITH HAND PLANES

Nowadays, the most popular method of filling in a frame is with a raised panel (a solid panel of wood with a flat center field that slopes to the edges at an angle). This is a clean and classic treatment with sharp corners and high relief that cast strong shadows, producing a very dramatic effect. With the availability of modern panel-cutting router bits, making raised panels for your doors isn't that difficult.

In a late-18th-century house I visited, I saw a pair of interior doors flanking an entry hall. The panels on these doors were an interesting and elegant variation on the raised panel. I call it the reeded and notched panel (see the photo at right).

The reeded panel, I believe, was devised by a country woodworker trying to elaborate on the flat panel. This treatment produces a panel surface that is entirely reeded, giving visual weight and emphasis to an otherwise plain field. The reeds can be laid either vertically or horizontally. It's a simple, but effective and unusual treatment of a panel.

It's rare to see reeds covering a large area, such as a panel. They are usually found grouped in sixes or eights along the length of a column or cut into short lengths and set into a molding or chair rail.

The panels on the doors were further decorated with notched corners. With the exception of the reeding plane, all of the work was performed with ordinary joiner's tools. Here's how.

REEDING THE PANELS

The work begins with the selection of a clear section of pine. Its surface has to be free of knots and other defects, with the grain running straight from one end to the other. After the panel is dimen-

The completed panel.

The moving fillister plane, with a depth stop, adjustable fence and nicker, can rabbet a panel all around to uniform dimensions.

A wide shoulder plane adjusts the thickness of the rabbet and removes any slight tearout left by the moving fillister plane. Because of its great weight and the low angle of the blade, this shoulder plane takes a fine shaving and leaves a perfect surface.

sioned for the frame, a moving fillister plane is used to rabbet the edges. A moving fillister plane is like a large rabbet plane, but it has a movable fence and depth stop, so the width and depth of the rabbet can be changed. It also has a nicker, or scribing iron. This small knife-like blade, set at 90° (see the top photo at left) precedes the plane iron and cuts or scores the fibers. On cross-grain cuts, it completely eliminates any tearout. After the panel has been rabbeted on all four sides, a wide shoulder plane is used to remove any traces of tearout and to adjust the thickness of the rabbeted edges (see the middle photo at left), ensuring an easy but snug fit into the frame.

The reeding is done with an unusual molding plane. Although the sole of the plane has been shaped for two reeds, this particular plane can only cut one reed at a time. The non-cutting reed on the outside is known as a follower (see the bottom photo at left), and it ensures uniform spacing between the reeds.

The reeding plane at left cuts only a single reed. The non-cutting reed alongside the blade rides piggy-back on the previously cut reed. The plane's direction can be reversed to remove any light tearout.

Since the plane doesn't have a fence, the first reed is the most difficult to cut, since the plane must be guided freehand along the left edge of the panel field. First the plane must be set for a light cut. The lighter the cut, the less the resistance to the cut and the more control you will have over the plane. The edge of the cutting bead is lined up with the edge of the panel field, and the plane is guided by the fingers of the left hand, which act as a makeshift fence (see the top photo at right). Once the cut is started, the plane will track by itself and cut the reed to the full depth. After the first reed is cut, the plane is moved over to the right. The follower is placed over the cut reed, and the next reed can be molded (see the middle photo at right).

MAKING THE NOTCHES

When the raised part of the panel is entirely reeded, the notched corners are laid out with a compass or a small template. The bulk of the waste is is cut off with a dovetail saw, then the outline of the curve is shaped with a carving gouge. The corner must be notched flush with the rabbeted edge—no deeper. Then the waste is carefully removed with a large flat chisel (see the bottom photo at right).

When cutting the first reed, the plane must be guided by hand and eye alone; subsequent passes can correct any minor error. Once the cut is established, the plane will track without difficulty.

The follower is placed over the cut reed in order to cut the next reed (right). This setup is repeated until the plane reaches the other edge of the panel field. Below, a wide chisel blends the notched corner perfectly into the rabbet.

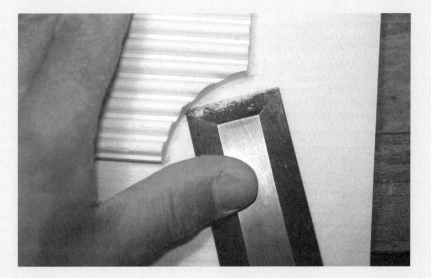

11

CORNER CUPBOARD

In the 18th century, owning pewter, silver, or china was an indication of wealth and social status, and these items required a proper display. What better showcase than a corner cupboard?

There's something special about a corner cupboard. The expected squareness and normal order of a room are breached by the imposing facade of this massive piece intruding into the room at an angle. It throws a room slightly out of whack, but also bestows on it importance and dignity, making the room seem larger than it is.

Built-in corner cupboards were always installed in a highly visible place—usually in the far right-hand corner of the right-hand room off the hall or entry. This location was almost always an exterior corner. Flanked by windows, the cupboard and its contents were flooded with natural light, illuminating the fine pewter and china on display.

Most historical examples extended to the ceiling and measured approximately 4 ft. across the front. Any broader, and a cupboard might be mistaken for an entry to another room; any narrower, and it wouldn't make an adequate impression. The best examples had fluted or molded pilasters, cornice moldings, and large doors with multiple lites (panes). Some cupboards were divided by a waist molding that was a continuation of the chair rail.

The corner cupboard was often part of an overall improvement or expansion of the original structure and was usually built and installed some time after the house was erected. Because they were so tall and wide, many corner cupboards were built in two parts, an upper case and a lower case. These separate cabinets, identical in plan but with different dimensions in elevation, were then stacked, and the resulting seam between the cases was camouflaged by the waist molding.

Most old timber-framed homes had massive corner posts that jutted into the room. In order to accommodate the intrusion of the corner post and still fit neatly against the walls, corner cupboards had their 90° back corner cut off. The outer corners were also cut away, but this was done to give the cupboard a cleaner facade and a dramatic projection off the walls.

The upper case was usually covered by a single glazed door, made up of anywhere from 9 to 20 separate panes of glass. These panes were set into rabbets and held in place with putty, in a fashion similar to window sash. The lower case was covered by a pair of frame-and-panel doors.

DESIGN

The corner cupboard in this project was built for the dining room of a beautiful 18th-century house in Warwick, New York. The overall dimensions of the cupboard were dictated by the ceiling height and the wall space to the left of a window near the corner.

The walls in this room were covered with antique wallpaper from the ceiling down to a chair rail, located 37 in. from the floor. The height of each case of the corner cupboard was determined by the position of the existing chair rail.

The cupboard I built for this house is shown in elevation in the drawing on p. 138; construction details and sections are shown later in the chapter, alongside the text that discusses them. But don't feel you must slavishly copy what I've done. Every room is different and calls for a treatment that enhances, rather than conflicts with, its proportions and mood. You might, for example, alter the outline of the top and bottom shelves to suit what you plan to display on them. I would suggest that you start by laying out a

CUPBOARD ELEVATION

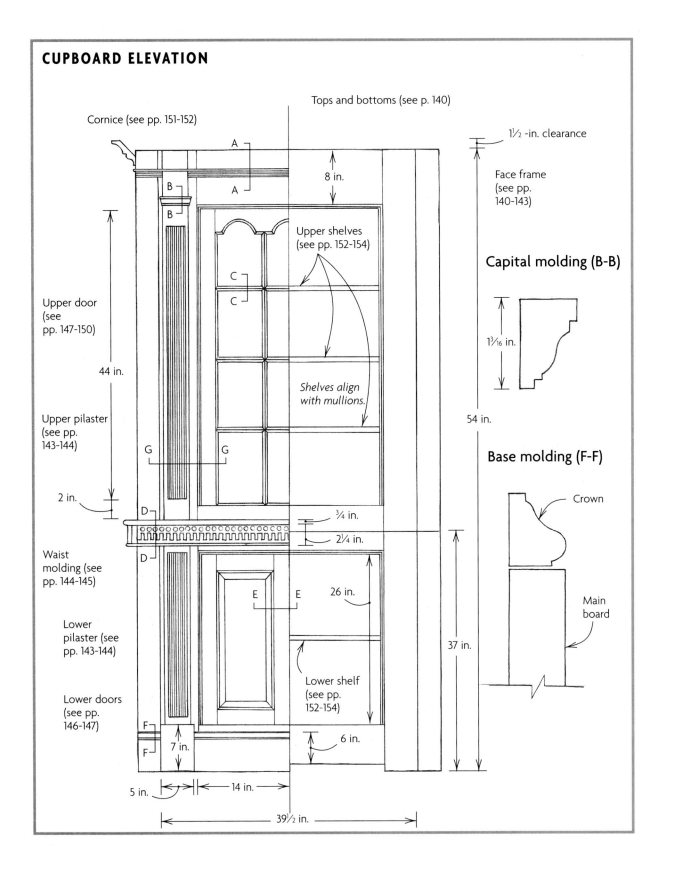

Cornice (see pp. 151-152)

Tops and bottoms (see p. 140)

A

A

8 in.

B

B

B

1½ -in. clearance

Face frame
(see pp.
140-143)

Upper shelves
(see pp. 152-154)

Upper door
(see
pp. 147-150)

C

C

44 in.

Shelves align
with mullions.

Upper pilaster
(see pp.
143-144)

G

G

2 in.

D

D

¾ in.

Waist
molding (see
pp. 144-145)

2¼ in.

Lower
pilaster (see
pp. 143-144)

E

E

26 in.

Lower shelf
(see pp.
152-154)

Lower doors
(see pp.
146-147)

F

F

7 in.

6 in.

37 in.

5 in.

14 in.

39½ in.

54 in.

Capital molding (B-B)

1³⁄₁₆ in.

Base molding (F-F)

Crown

Main
board

full-size plan to determine just how the cupboard will fit into the room. See first if it will accommodate the flow of traffic or interfere with it. From there, go on to design the details.

As you design, keep in mind that the combined height of the two cases should be at least 1½ in. less than the floor-to-ceiling dimension of the room. That way you will be able to level, plumb, and secure the lower case, then lift the upper case easily into place without damaging the ceiling or walls. The gap at the ceiling will be concealed by the cornice molding.

MATERIALS

For the carcases, use ¾-in. birch shop-grade or cabinet-grade plywood. It will cost a little more than a utility-grade plywood, but it's smoother and less likely to warp. Later, the edges of the plywood will be covered by the face frame, which is 1-in. solid pine, as are the doors. The cornice molding is ¾-in. pine.

In period examples, the sides and back of the carcases were usually made up of narrow tongue-and-groove boards, but using plywood will save you a lot of work because you can cut these large sections out of a 4x8 sheet, greatly simplifying the construction.

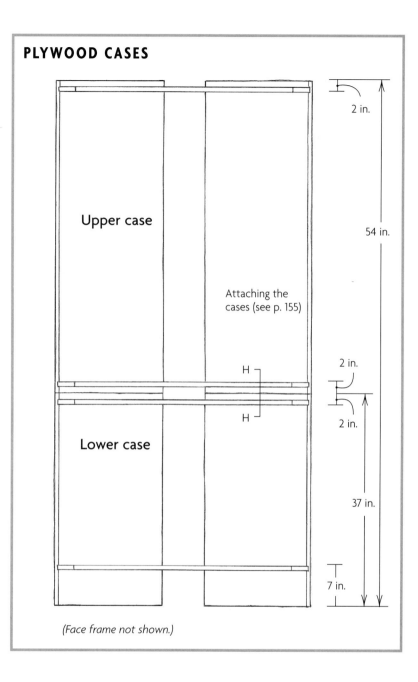

PLYWOOD CASES

Upper case

Attaching the cases (see p. 155)

Lower case

2 in.

54 in.

2 in.

2 in.

37 in.

7 in.

(Face frame not shown.)

For this cupboard, I purchased about 32 board feet of ¾-in. pine and about 40 board feet of 5/4 pine. These are generous quantities, but there are advantages to having more wood than you really need. You can pick and choose, using only the best material contained in the boards, and you also have some extra in case you make a mistake. The choicest material should always be reserved for door parts.

SIDES, TOPS, AND BOTTOMS

Cut out the sides of both cases (see the drawing on p. 139), then cut ¾-in.-wide dadoes for the tops and bottoms of these cases. On the upper case, make the dadoes 1¼ in. from both ends; on the lower case, make the top dado 1¼ in. from the end and the bottom dado 6 in. from the end. Where the cases meet, blocking will provide the means of attaching the upper case to the lower case (see the drawing on p. 155).

Each case has a top and a bottom. Since this is a corner cupboard, the tops and bottoms of the carcase, all four of them, are triangular with the corners cut off. The outline of the top and bottom pieces defines the shape of the cupboard, so cutting them might seem a little tricky. If you follow the dimensions and angles detailed in the drawing at right or in the one you made for your own design, everything should turn out fine. Be sure to allow for the depth of the dadoes. Use the piece you cut as a full-size pattern for the remaining ones.

For the tops and bottoms in my corner cabinet, I ripped a length of plywood to width (21¾ in.), then cut the 135° side and back-corner angles freehand on the

bandsaw. Corner cuts that aren't straight can be dressed on the jointer or with a bench plane. When all the pieces are cut, stack them and check for discrepancies; trim where necessary with a block plane.

THE FACE FRAME

A face frame wraps around the upper and lower cases at the frontmost corners. The angle of this mitered corner is 135°, so each half of the miter will have to be cut at 67½°. Since these pieces are wide and the miter joint will be visible in the finished piece, you should employ some mechanical means to keep the seams tight and gap free.

There are two methods I recommend. You can groove both edges of the joint for a spline (see the photo on the facing page), then glue these parts together before attaching the face-frame parts to the carcase. Alternatively, the joints between the stiles and rails of the face frames can be executed using biscuits.

With the spline method, start with the lower case, which is smaller, and place a mitered stile on the plywood carcase. It should lie flat and in full contact at the angle of the plywood carcase. Next, attach the rails, followed by the remaining stile. Apply glue to the face-frame

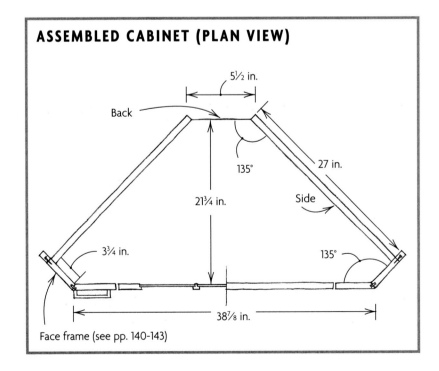

ASSEMBLED CABINET (PLAN VIEW)

Back

5½ in.

135°

27 in.

Side

21¾ in.

3¾ in.

135°

38⅞ in.

Face frame (see pp. 140-143)

A splined miter joins the two parts of the face-frame corner stile.

parts and attach them to the carcase with narrow-crown staples or finishing nails.

Since the interior of the cupboard is often painted a different color than the outside, I cut a shallow rabbet along the inside top-edge of the lower carcase rails (for both upper and lower halves of the cupboard) to act as a paint break.

In the 18th century, the woodwork would have been installed before the application of plaster, eliminating the need for scribing. Today, a cupboard like this is installed after the walls are finished and must be fitted to accommodate the undulations of the wall. You could make the frame 1 in. wider and scribe that

(text continues on p. 143)

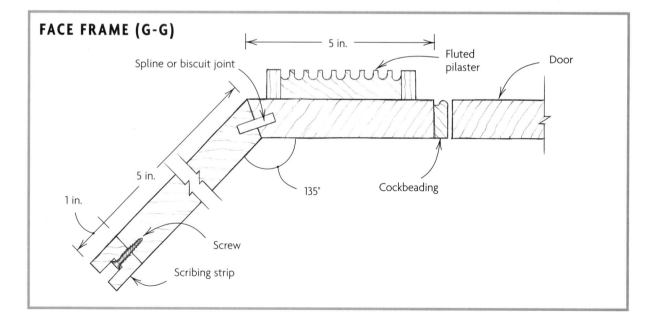

FACE FRAME (G-G)

5 in.

Spline or biscuit joint

Fluted pilaster

Door

5 in.

1 in.

135°

Cockbeading

Screw

Scribing strip

FLUTING COLUMNS WITH A MOLDING PLANE

Nowadays, in a large cabinet or millwork shop, fluted material is often produced on machines with cutters that can cut multiple flutes of equal width, spacing, and depth in a single pass (see the photo below).

A small-shop woodworker who needs a pair of fluted columns for a fireplace mantel or front door can turn to the router or shaper equipped with a single-flute cutter. After each pass, however, the fence has to be moved to place the next flute in the proper location, and the setup must be checked with test workpieces. The constant measuring and double-checking are at best tedious.

In the 18th century, furniture makers and carpenters relied on molding planes to produce all their millwork. And there were planes that could cut multiple flutes in a single pass. But if a particular molding plane produced a $1/2$-in.-wide flute, the actual cutting width was the combined arcs of all the flutes. Imagine the effort needed to produce a profile $1\frac{1}{2}$-in. wide. If a pair of flat fluted columns was needed, it was easier to use a single-flute plane (see the top photo on the facing page) and cut the flutes, one at a time, using nothing more than the hand and eye as a guide for positioning and spacing.

The non-cutting portion of the molding plane's sole is the stop. When the plane begins cutting the profile, the stop is raised off the wood surface. With each pass, more of the profile is cut and the stop comes closer to the wood surface. When the flute is fully realized, the stop will ride on the surface of the workpiece, preventing the plane from cutting any deeper. This integrated stop ensures that all the flutes will be uniform in width and depth.

The width of the stop is used to set the position of the first flute from the left-hand edge of the board. With your fingertips riding along the edge of the board (see the top photo on the facing page), carefully guide the plane down the length of the board. Since the plane initially takes such a light shaving, any meandering cuts can be redirected and straightened in subsequent passes. When the developing flute looks straight and parallel to the edge, simply let the plane ride in the flute until the full flute is formed. Then again, using the width of the stop for spacing, position the plane alongside the previously cut flute. This time, guide the plane with your pinky riding in the previously cut flute (see the middle photo on the facing page). Repeat the sequence for the rest of the flutes (see the bottom photo on the facing page). Any errors in spacing or straightness can be spotted only if someone sights down the length of the molding. Once it's up on the wall or the cupboard, no one will notice.

Shaper and molder cutters can produce multiple flutes in a single pass.

In the 18th century, woodworkers relied on molding planes to produce uniform and evenly spaced flutes. At left: A single-flute molding plane at the edge of a board. The fingers of the user's left hand act as a fence, maintaining the plane's position as it cuts the first flute. Below and bottom: The pinky of the left hand runs along the inside of the previously cut flute, placing the second and subsequent flutes parallel to the first.

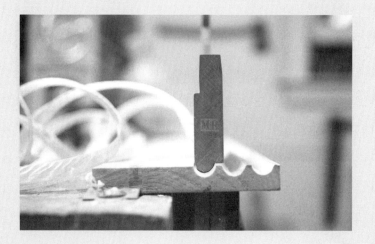

to fit, but that would be an extremely cumbersome operation on a piece this size. Instead, my design calls for adding 1-in. removable scribing strips to the frame (see the drawing on p. 141). You attach the strips, scribe the cabinet to the wall, and then remove the strips for easier trimming. When the strips have been cut to the line, reattach them to the cupboard. The strips should be secured with countersunk screws.

The face-frame stiles are finished off with cockbeading on their inside edges. With the face frames attached to both cases, run off about 22 ft. of cockbeading using a molding plane, a router bit, or a scratch stock. Cut the molding free from the board and attach it to the inside edges of the stiles and the top rails with glue and finishing nails.

THE PILASTERS

Fluted pilasters applied to the face frame flank the doors of the upper and lower cases. You can cut the flutes by hand with a molding plane (see the sidebar on the facing page) or on a router table or shaper; you can also purchase ready-made fluting by the linear foot at most home centers or any lumberyard. The material used for this project was purchased from a millwork shop. Each flute is ¼ in. wide,

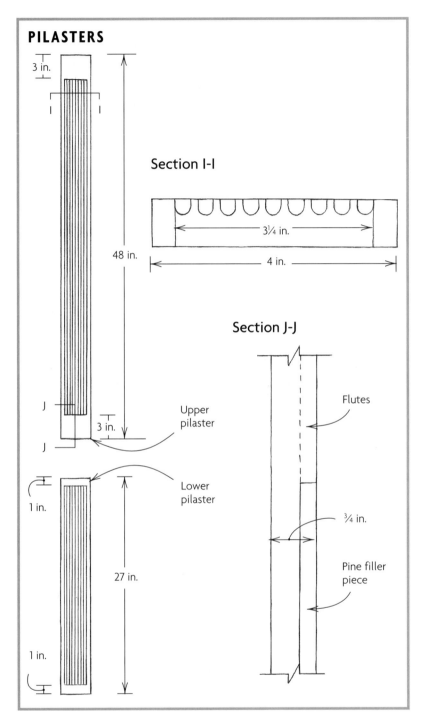

PILASTERS

3 in.

Section I-I

3¼ in.

4 in.

48 in.

I   I

J
3 in.
J

1 in.

27 in.

1 in.

Upper
pilaster

Lower
pilaster

Section J-J

Flutes

¾ in.

Pine filler
piece

with a ⅛-in. flat between them. There is no strict rule regarding the size or spacing of the flutes. Let your tastes and the design of your cupboard dictate the dimensions.

Unless you jig up to produce stopped flutes, you should plan on framing the fluted material. Framing the flutes gives the pilasters a clean and contained appearance, which is preferable to letting the flutes die into the plinth or run behind a molding (see the drawing at left).

At each end of the pilaster strip, cut a shallow rabbet. Then, fill in the rabbet with a thin piece of pine (see the photo on the facing page). On the sides of the pilasters, along the full length, glue and clamp strips of pine to complete the frame. When the glue is dry, plane everything flush. Be careful to remove any machine marks left by the table saw on the pilasters (see p. 30).

THE WAIST MOLDING

The waist molding is an important part of this corner cupboard. Unlike the cornice, which runs along the top, the waist molding is down low, below eye level, where it can be examined and touched. It is one of the things that give this cupboard its distinct character.

I wanted something light that suggested lace, a pattern that would be easy to produce on shop machinery, yet have a hand-carved look. People usually take a few minutes to figure out how it was made—some never do.

The pattern (see the drawing below) is a variation on dentil molding. Instead of square and severe cutouts, it has softer, angled crenelations. These ½-in. cutouts exactly ½ in. apart are cut using a jig similar to the

jig that I used to cut the checkerboard fireplace mantel (see pp. 100-102). This jig, however, has a bed set at an angle to produce this interesting effect (see the top left photo on p. 146). After the angled dentil cuts were made, I drilled a clean ¼-in.-dia. hole, above each recess (see the top right photo on p. 146).

The full waist molding is composed of four parts: the foundation strip, the upper beaded shelf, the crenelated band, and the lower beaded shelf. When all the parts are ready and sanded, glue and clamp all but the crenelated band together into a single unit, then attach it to the cupboard with with #8 screws (see the bottom photo on p. 146). When the band is applied later, it will cover the screws.

The rabbeted ends of the fluted sections are fitted with a pine filler piece as the first step in framing the pilasters.

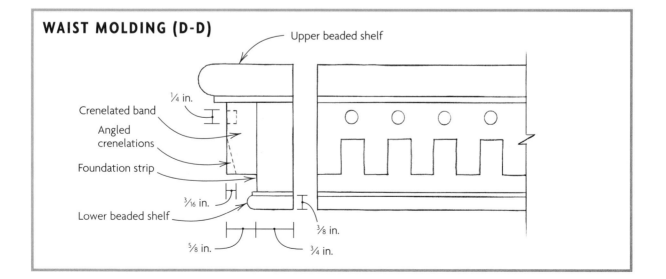

WAIST MOLDING (D-D)

Upper beaded shelf

Crenelated band

Angled crenelations

Foundation strip

¼ in.

Lower beaded shelf

³⁄₁₆ in.

⅝ in.

¾ in.

⅜ in.

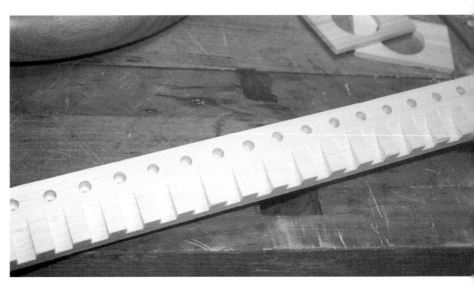

A jig with an angled bed is used to cut the angled crenelations for the waist molding (left). Holes drilled above the angled cuts complete the pattern (above).

The base of the waist molding is screwed to the cupboard. The screws will be covered by the crenelated band.

THE LOWER DOORS

The lower doors are made of solid pine with a flat rabbeted panel. The stiles and rails of the door frames are joined with mortise-and-tenon joints.

To build the door frames, prepare your stock to the dimensions shown in the drawing on the facing page, then cut the door parts to length (be sure to add the length of the tenons to the rails. Next, cut a ¼-in.-wide, 1-in.-deep groove along the inside edge of the rails and stiles. This can be easily done on the table saw using a 40-tooth combination blade and raising the blade ¼ in. with each successive pass.

LOWER DOORS (E-E)

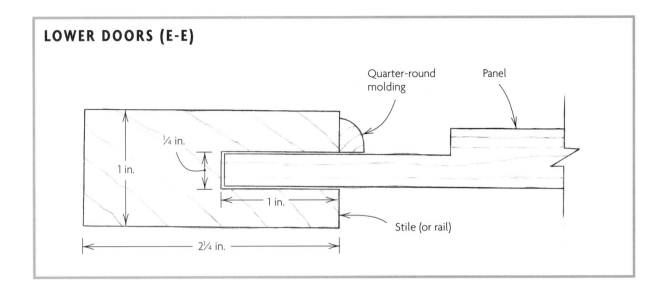

Quarter-round molding

Panel

¼ in.

1 in.

1 in.

2¼ in.

Stile (or rail)

On the ends of the rails, cut simple stub tenons that will fit into the grooves in the stiles. Assemble the frames without glue to check dimensions, then measure for the panels. Cut the panels slightly smaller than the full depth of the grooves so they will have room to expand in the frame.

Rabbet the panels all around to fit into the groove on the stiles and rails. Be sure to sand all the parts before assembly and pay particular attention to inside edges and surfaces that won't be accessible after assembly. Once each door is dry-fitted, apply glue only to the frame parts, not to the panel, which must remain free to move within the frame.

When the door frames are dry, attach a small quarter-round molding along the inside edge of the frame and against the panel. Be sure to secure the molding only to the frame and not to the panel.

THE UPPER DOOR

The upper door is a lot like a window sash. The stiles and rails form an outer frame that supports the glass. The stiles run the full height of the door, with the rails coped into them. The interior of the door frame is divided by mullions, running horizontally and coped into the stiles. The shorter muntins run vertically and are coped into the mullions.

Typically, good period examples have anywhere from 9 to 20 panes, or lites. Naturally, the number of lites will depend on the overall size of the door, but keep in mind that earlier cupboards had smaller lites (typically 6 in. by 8 in.) while on later examples lites might be as large as 10 in. by 12 in. This door has 12 lites.

A sash door with 12 lites, each measuring almost 8 in. by 10 in., would overpower the cupboard. If the sash had been left square, the door would look like a store window. Instead, to animate the design and create additional interest, the three top lites are arched in a tombstone pattern. This introduces some curves and some lightness to the design and directs the eye upward to the cornice.

Layout

In period examples, the parts of the sash-door grid are molded with a plane, then carefully mortised and tenoned together. This is meticulous work, and any small error in your layout or calculations can cause a misplaced muntin or mullion, giving the sash a lopsided appearance and ruining the door.

To lay out this door, I used story poles (one for the vertical axis of the sash, another for the horizontal) to mark the exact positions of the muntins and mullions on the stiles and rails. On these poles, I traced the exact profiles and dimensions of the sash components, which I simply measured off later when cutting the parts to length.

Molding the stock

Instead of coping and mortising the door parts together. I used a cope-and-stick router-bit set that is modified for sash. This matched pair of ½-in.-shank router bits produced the molded edge and the rabbet for the glass with one cutter (see the drawing below). The other bit produces the coped profile that interlocks precisely with the molded edge and rabbet (see the bottom photo on the facing page).

After ripping the stock for the muntins and mullions, I molded both edges of each strip on a router table fitted with a fence. Then, using my story poles as a guide, I cut the individual pieces to length, changed router bits, and routed the end of each rail, muntin, and mullion with the coped profile. To support the thin pieces safely and ensure accuracy as they are run past the router bit, you can use a shop-made jig fitted with a backup piece to reduce tearout on the exit side of the cut.

Cope-and-stick is not the usual way period sash doors were made, but it has a big advantage

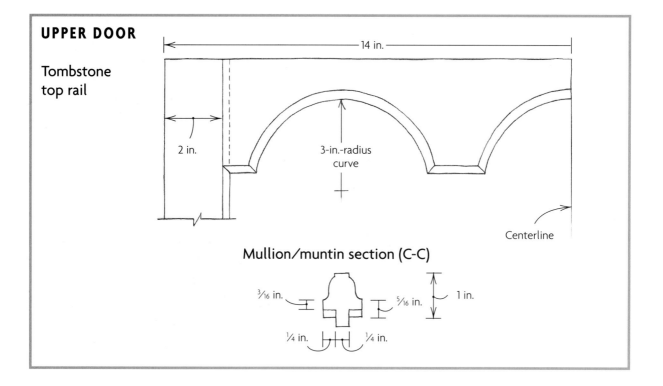

UPPER DOOR

Tombstone top rail

14 in.

2 in.

3-in.-radius curve

Centerline

Mullion/muntin section (C-C)

3/16 in.   5/16 in.   1 in.

1/4 in.   1/4 in.

Sash joists that are overlapped and scribed, as was done in the 18th century, must be made perfectly—their position cannot be adjusted.

over the traditional method. If the joints had been scribed and overlapped or done in mortise and tenon in the traditional 18th-century fashion, it would have been impossible to make any adjustments in their position within the sash frame (see the photo above). Since the joints are only coped, the ends can easily be repositioned along their reciprocal part to achieve perfect spacing and uniform pane size. The sash might not be as strong as one made with traditional methods, but once the glass is held in place with glazing pins and puttied, it will be strong enough.

Sash joints that are coped but not mortised can easily be adjusted to achieve perfect spacing within the sash frame.

The upper sash rail

The tombstone cutouts can be made on the bandsaw. On my cupboard, once I established the radius of the curve (see the drawing on p. 148), I marked it directly onto the rail, then cut to the line freehand and cleaned up the cuts with a spokeshave and file.

The next step, molding the upper rail, is a little more difficult, but you can do it on the router table if you replace the fence with a pivot pin. A pivot point is a post set into the table surface about 3 in. from the center of the router bit. The pin supports one end of the workpiece while you slowly pivot the rest of it into the bit. If a workpiece is pushed into a rotating bit without the support of a fence or pivot pin, the bit might grab the piece from the operator and fling it back at the operator. A pivot pin lets you start the cut safely and ensures good results.

Assembly and glazing

When all the stiles, rails, muntins, and mullions are grooved, molded, coped, and cut to length, verify the dimensions of each piece, either by measuring or by laying each piece directly onto your story pole. If everything checks out, dry-fit the sash together, then glue and clamp.

The cupboard door has 12 lites. For these you could salvage antique panes from discarded window sash, but the glass you get will probably not be uniform in terms of waviness (distortion), color, and seeding (the presence of small air pockets). The differences from lite to lite will be distracting and detract from the look of the finished piece. That's why I recommend restoration glass, which you can get from S. A. Bendheim, at 120 Hudson St., New York, NY 10013.

Reproduction glass is available in different degrees of waviness. Since this glass is outrageously expensive, I let a professional cut it and glaze the sash. All I had to do was provide the glazer with a template for the tombstone lites.

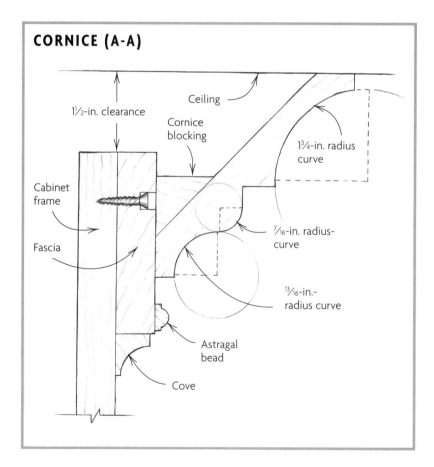

CORNICE (A-A)

Ceiling

1½-in. clearance

Cornice blocking

1¾-in. radius curve

Cabinet frame

7/16-in. radius-curve

13/16-in.-radius curve

Fascia

Astragal bead

Cove

THE CORNICE MOLDING

The design of a cornice molding can make or break a piece. For this cornice molding, I wanted something with good projection at the top and a small tight drape at the bottom. The upper portion would give the cornice some lift and upward movement, and the bottom would look like fabric unfurling, catching light and creating shadow. Since there was nothing like that available in stock moldings, I made my own. Its profile is shown in the drawing on the facing page.

Producing this molding requires the execution of three machine operations: two cove cuts and one V-cut; the sequence is shown in the drawing at right. The wide cove cuts on this molding can be made on the table saw. This is a basic table-saw technique that is really simple and almost foolproof. It involves running the stock over the table-saw blade at an angle, along a fence. Instead of cutting straight through, the blade will scrape the underside of the workpiece, producing a cut that corresponds to the elliptical arc of the blade. After each pass, the blade is raised a little until the desired depth and width of cut are obtained. The greater the angle of approach to the blade, the wider and flatter the cove. At 90° to the blade,

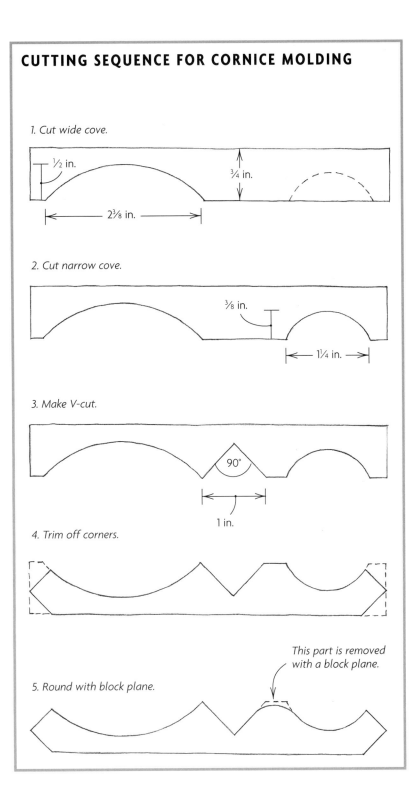

CUTTING SEQUENCE FOR CORNICE MOLDING

1. Cut wide cove.

½ in.   ¾ in.   2⅜ in.

2. Cut narrow cove.

⅜ in.   1¼ in.

3. Make V-cut.

90°   1 in.

4. Trim off corners.

5. Round with block plane.

This part is removed with a block plane.

Any machine marks left on the cornice molding are removed with sandpaper backed by a sanding block.

the cove. At 90° to the blade, the elliptical cove flattens into a circular arc.

Begin by plotting out the profile of the molding on the ends of your molding stock. Raise the blade to the full height necessary to achieve the depth of cut, then angle the stock to the blade and align the outline of the first cove cut with the perimeter of the blade. Next, place a fence along the angled molding and clamp it securely to the table saw. Now lower the blade until it projects no more than ⅛ in. from the table. Turn on the saw, and slowly and carefully, pass the stock over the spinning blade.

The first pass will leave a tiny cove that should be centered on your outline. After each pass, raise the blade a little to widen and deepen the cove cut until you've cut to your outline. As the blade is raised, resistance to the cut will increase. So in order to maintain safe control over the operation, take lighter cuts as you approach the full cove cut, and use a push stick.

Next, reposition the fence for your narrow cove and repeat the operation. When both coves are cut, set the blade at 45° and cut the V. With the blade still angled at 45°, cut both edges of the

molding so it will meet the cupboard and the ceiling at the proper angle.

When all these cuts are made, round off the flat between the V-cut and the smaller cove with a block plane. Then sand the molding to remove saw marks and any ridges or flats left by the plane (see the photo above).

THE SHELVES

There are four shelves in this cupboard: three in the upper case and one in the lower (see the drawing on the facing page).

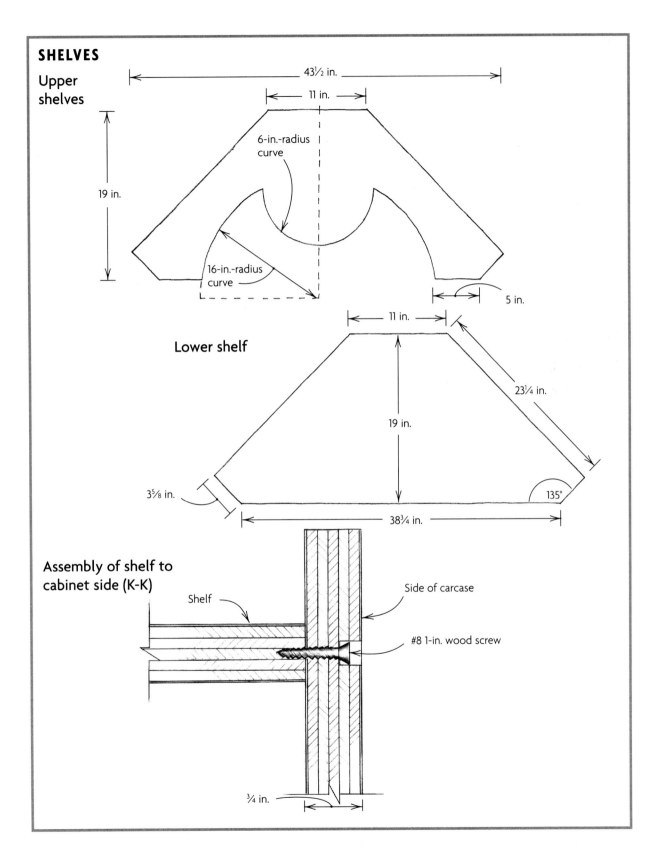

SHELVES

Upper shelves

43½ in.

11 in.

6-in.-radius curve

19 in.

16-in.-radius curve

5 in.

Lower shelf

11 in.

23¼ in.

19 in.

135°

3⅝ in.

38¾ in.

Assembly of shelf to cabinet side (K-K)

Shelf

Side of carcase

#8 1-in. wood screw

¾ in.

Exposed plywood edges are covered by $^{1}/_{16}$-in. edging, glued and held in place with masking tape. When the glue is dry, the tape will be removed and the edging trimmed flush with the surface of the shelf.

The upper-case shelves are fixed in line with the mullions that cross the sash door horizontally. They can be dadoed into the sides of the carcase and attached with screws, as shown in the drawing on p. 153, or they can be supported by cleats attached to the interior walls of the upper unit.

The shape of these shelves is designed for displaying flat china or pewter along the back of the cabinet while creating a prominent circular perch at the center of the shelf for a large bowl or vase.

The outline of the shelves can be cut out on the bandsaw or with a hand-held jigsaw, then cleaned up with files and sandpaper. The curved edges of the plywood can be covered with thin edging, glued and held in place with masking tape (see the photo above). When the glue dries, trim the edging flush with a cabinet scraper, then sand.

The lower-case shelf extends to the cabinet doors. This shelf has a straight front edge, which can be banded just like the upper shelves.

ASSEMBLY

You can assemble the two cases, install them in place, then attach the columns, moldings, and doors. Or you can put the whole thing together in the shop, minus the base, waist, and cornice moldings. Each approach has its merits. Assembling the cupboard in the shop allows you to fit all the parts together precisely. Any gaps or problems between the components can be fixed quickly in the proper manner; doors can be neatly fitted and hardware precisely placed. Solutions to problems encoun-

ATTACHING THE CASES (H-H)

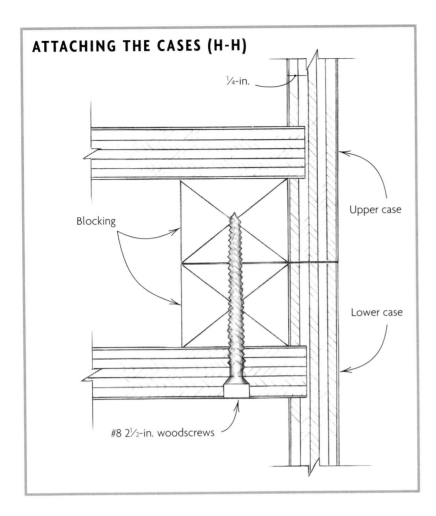

¼-in.

Upper case

Blocking

Lower case

#8 2½-in. woodscrews

are fit from the inside of the carcase. When a good fit has been achieved, glue and nail them into place, and apply the face frames, using one of the methods described on pgs. 141 and 143.

Attaching the pilasters

Many 18th-century cupboards have the pilasters on the side of the face frame, close to the wall. On this cupboard, however, I placed the pilasters on the front of the face frame, close to the doors. Here they give the cupboard a more dramatic facade, but complicate the application of the cornice molding. When the pilasters are applied to the side of the face frame, there is room for the required series of miters and small mitered returns. When the pilasters are placed on the front of the cupboard, close to the projecting corner of the cupboard, there isn't enough room to execute the required miters and returns in an attractive way. To circumvent the problem, the end of the pilaster runs into a fascia board that spans the cupboard face frame (see the photo on p. 156).

The fascia and the pilasters are secured to the cases with #8 screws. You can screw through the front of the pilaster wherever molding will later cover the exposed screw. If no molding would go over the screw, attach the pilaster to the case from the inside.

tered during installation are often poor improvisations. However, assembling the cupboard on site allows you to solve unanticipated problems, should any arise.

Assembling the cases and face frames

After applying glue to the joints, assemble the cases using #8 1-in. woodscrews, countersunk and spaced 5 in. apart (see the drawing above) to hold the parts together. Check the cabinets for square. The two carcases are large, and their odd shape makes them difficult to move around and nearly impossible to clamp up. Usually, if the screws holding the case together are driven in squarely, the carcase will remain square. If the cabinets shift out of square, brace them with 1x2 strips nailed diagonally across the face.

When the case is assembled, measure for the two back panels. Both have edges cut to 45° and

The pilaster, topped by a block, continues upward into the fascia along the top of the cupboard.

Between the fascia and the fluted pilaster there is a separate block attached to the carcase. It is meant to appear as a continuation of the fluted pilaster. The block itself can be left plain, as I decided to do here, or it can be adorned with a carved rosette. The piece can be attached from the front of the cupboard with 4d finishing nails or screwed from inside the cupboard.

Putting up the moldings

Once the fascia board and the pilasters have been attached, the ⅜-in. astragal bead and the ¾-in. cove can be mitered and nailed in place. The cuts can be made either by hand or with a power miter saw. Either way, aim for clean cuts and tight miters. Don't accept less and resort to wood fillers later.

Small moldings should be attached carefully. For best results, I recommend using a pneumatic brad nailer or pin nailer (see p. 106). The smallest pieces should be glued, instead of nailed, and held in place with masking tape until the glue dries.

Once the moldings have been attached, fill and sand any small gaps, nail holes, and faults. I use Elmer's water-based wood filler. This putty-like material dries quickly, with minimal shrinkage, and is easily sanded flush.

Installing the doors and shelves

Plane the doors to fit the openings, allowing a full ⅛-in. reveal. This generous allowance will be taken up by the primer and the finish coats of paint, as discussed at the end of this chapter.

The next step is installing the hinges. Surface-mounted H hinges are the easiest to install, and you can antique them first if you like (see p. 35). First, mount one leaf of the hinge onto the door, with the barrel and the other leaf hanging clear. Then place each door into its opening with uniform clearance all around, using small wedge-shaped shims to adjust the spacing and support the door in place. When the door is positioned, predrill and attach the remaining leaf to the face frame.

The three upper shelves must align with the mullions of the sash door. So after the sash door is installed, mark the location of the mullions onto the vertical edges of the face frame, then remove the door. Lay the cupboard on its side, align the shelves with the marks on the face frame, then screw the shelves in place through the sides. The single shelf for the lower half of the cupboard can be placed anywhere you like, since it will be covered by a solid door.

INSTALLATION

Position the lower case in the corner and check it for level and plumb. If necessary, trim material off the bottom to achieve solid support for the full cabinet. Next, scribe the add-on strips to the wall (see p. 143). If only a small amount of wood needs to be removed, do it with the strips attached. Otherwise, remove the strips and bandsaw just shy of the scribe line, then reattach the strips. Small gaps, (about 1/16 in.) can be caulked.

Once the bottom case has been scribed and positioned snugly against the wall, slip shims between the wall and the back of the cabinet, filling any gaps (see p. 90 for a similar technique). Then attach the case to the wall with #8 3½-in. screws driven through the case sides into the wall. Now rehang the doors and check for square. Repeat the same steps for the upper carcase. When both halves of the cupboard are in place, screw them together through the blocking (see the drawing on p. 155). Two 2½-in. screws on each side should be sufficient.

The last step is to attach the base molding, the remaining waist molding pieces, and the entire cornice.

FINISHING

Finishing the cupboard is a critical part of this job. I've seen delicate detail almost obliterated by heavy, viscous paint applied with a cheap brush, turning a woodworking masterpiece into a globby, dripping disaster. Conversely, a good paint job can really cover a multitude of sins.

A simple routine I follow involves the application of a good-quality water-based primer. After sanding the primed surface, I fill any small nicks or dents with wood filler or joint compound. The last step before painting is to caulk all the seams. This cupboard was finished with three thinned coats of satin-gloss latex paint.

12

MAILBOX POST

Consider the humble mailbox post. Usually it's just a 4x4 stuck into the ground or a bland blob of molded plastic—hardly the sort of thing you expect to find in front of a traditional home. Yet a mailbox can be a thing of beauty. The one you see in the photo on the facing page combines timber-frame joinery with gate-post trim—a post cap, beaded molding around the post, and a finial. It's a small-scale project that doesn't call for a large outlay of time or money, and it also happens to be a design that will look good in front of a Colonial saltbox, Craftsman-style bungalow, or even a Tudor castle.

The drawing at right shows the parts and overall dimensions of this mailbox post, but don't rely on it blindly for your own project. First go out and buy the mailbox, because its size and design might bear on the post dimensions. Also check at your local post office for the requirements concerning height and delivery access. When I built a beautiful new mailbox for my own house, I didn't do this, and later I learned from the postmaster that I had to plant it across the road in front of my neighbor's house because mail was not delivered to my side of the road. So I widened the shelf to accommodate my neighbor's

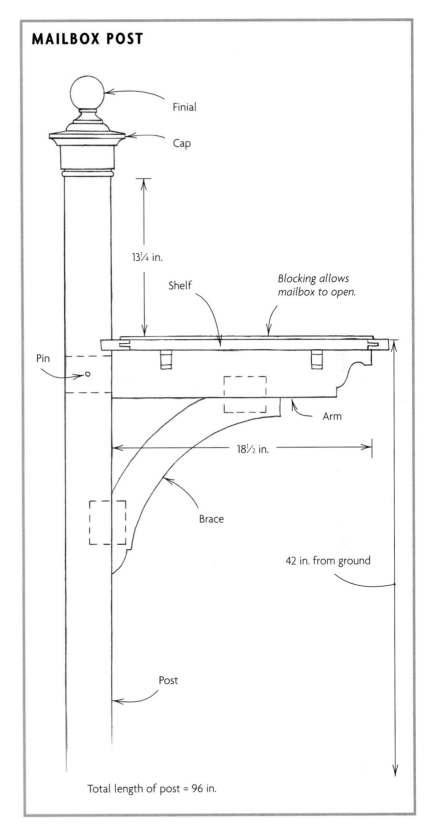

MAILBOX POST

Finial

Cap

13¼ in.

Shelf

Blocking allows mailbox to open.

Pin

Arm

18½ in.

Brace

42 in. from ground

Post

Total length of post = 96 in.

mailbox as well as my own. Needless to say, that wasn't exactly what I had in mind.

MATERIALS

For this project, redwood is your best bet. It's easy to machine and work with hand tools, and it looks good and weathers well. At the lumberyard, pick through piles and bins to find several quartersawn 4x4s with dead-straight grain on all four sides. For this project you will really need only two 8-ft. 4x4s, but I recommend buying several. For this project you will want to cut off any damaged portions and use only clear, straight stock. Buying an extra 4x4 or two now makes more sense than having to make another trip to the lumberyard later. Good and bad pieces are usually all the same price, so take your time and pick out the best. Quartersawn stock is attractive, easy to plane, and will remain straight. Also pull some 5/4 stock for the mailbox shelf and post cap, and some ¾-in. stock for the post-cap moldings and mailbox shelf.

If you have a lathe, you can turn the finial, but if you don't, just choose something you like at a lumberyard or home center.

You can purchase reasonably priced ready-made post caps and finials in cedar, redwood, or pressure-treated fir. For this post, I selected a simple round finial.

STOCK PREPARATION

The redwood that you get at lumberyards comes with the edges eased, since it is generally used for decks. This project calls for crisp, clean edges and smooth sides, so most of the wood will need to be thickness-planed or ripped. Take the post and arm down to 3¼ in. and the mailbox shelf to 1 in. You can leave the shelf and molding stock at ¾-in. thickness; just trim ¼ in. off the edges to square up the stock. The brace is 2½ in. thick.

THE POST

The post should be left as long as possible to ensure adequate support when one end is buried in the ground; it's a lot easier to cut off any excess later than to have to splice on additional wood. Choose the best end of the post to show at the top. Measure down 14 in. from the top and mark out the thickness of the arm across the post. These lines mark the exact position of the arm so the mortise can be laid out.

The usual guideline is to make the mortise and tenon one-third the thickness of the mortised piece. But because redwood is relatively soft, a 1-in. tenon seems a little thin; a 1½-in.-wide tenon, with ⅞ in. of wood on either side of the mortise, would be much stronger (see the drawing on the facing page). Considering the extra strength and support the curved brace gives the post, I think these dimensions work out fine.

A perfectly cut mortise-and-tenon joint starts with layout. Use a marking gauge or a square and a marking knife to lay out the mortise on both sides of the post. Never rely on a pencil line, which might smudge or become fuzzy. A scribed line is clear and unmistakable.

Next, carefully drill out the waste on a drill press using a ¾-in. Forstner bit or a sawtooth bit (see the photo on the facing page). Forstner bits have a smooth rim, and sawtooth bits have jagged teeth. I find that sawtooth bits work better in soft-wood in thicker dimensions, but you can safely use either type of bit. Don't try to drill all the way through at one time; go about ¾ in. deep, then back out the bit to clear the hole of chips before going deeper. By carefully placing the center of the bit for each hole, drilling the holes slowly, and clearing them frequently,

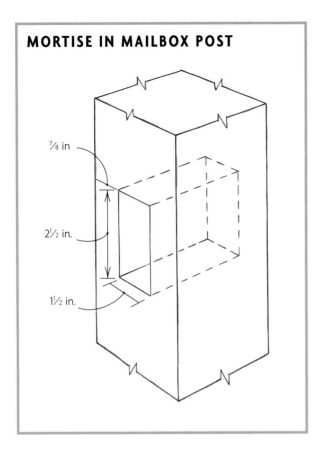

MORTISE IN MAILBOX POST

⅞ in

2½ in.

1½ in.

A Forstner bit in a drill press roughs out the mortise in the post. Make shallow cuts and withdraw the bit frequently to keep the hole clear of chips.

you can obtain clean straight holes contained within the mortise outline.

With the bulk of the waste removed, pare to the lines using a 1½-in.-wide paring chisel. This tool has the weight and width necessary to produce broad, flat surfaces within the mortise. Pare into the mortise from both sides. When you reach the layout line, look into the mortise, sighting down along the edge. Anything sticking out will interfere with a perfect fit. Finish the preliminary paring with a 1-in.-wide chisel, fine-cut rasp, or a planemaker's

float. To avoid tearout and damage to the visible surfaces of the joint, work from the outside into the joint.

THE ARM

The arm (see the drawing on p. 162) is 22 in. long (including the tenon) and the same thickness as the post. It will have a decorative scroll at one end and two dadoes on top to receive the brackets that support the mailbox shelf. But the first task is cutting the tenon and fitting it to the mortise in the post.

After marking out the joint completely, cut the shoulders of the tenon on the table saw. Stay just shy of the layout line, keeping them a little fat. Next, cut the cheeks. It's a lot safer to make this cut on the bandsaw than on the table saw. A bandsaw equipped with a ½-in., 3-tpi blade with pronounced hook-style teeth will deliver a clean and straight cut. Set the blade just outside of the line and push the arm slowly. Cutting the tenon a little fat gives a margin to correct any drift in the blade.

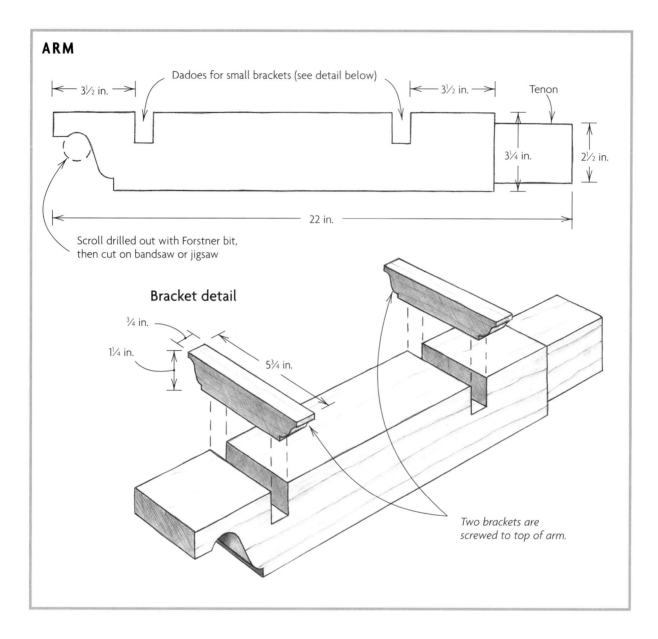

ARM

Dadoes for small brackets (see detail below)

3½ in.

3½ in.

Tenon

3¼ in.

2½ in.

22 in.

Scroll drilled out with Forstner bit, then cut on bandsaw or jigsaw

Bracket detail

¾ in.

1¼ in.

5¾ in.

Two brackets are screwed to top of arm.

If the cut is guided carefully and slowly, blade drift shouldn't be a problem.

Now secure the mortised post on a bench or in a vise. Then carefully place the tenon on the arm up against the mortise and visually check the fit. Don't bang

anything together. Just see how close the fit is and by how much the tenon needs to be trimmed. Resist the temptation to return to the bandsaw. You might saw off too much and then have to repair or shim the tenon later. Instead, resort to planes.

On my own mailbox post, I used a shoulder plane to trim the tenon. I own two of these planes, and both are antique English planes. These planes are dovetailed of mild plate steel, instead of cast, and are filled with rosewood. Their mouths are very small, permitting only fine shavings to pass. Because of their weight, they don't chatter or bounce around—even on cross-grain hardwood cuts. And their blades extend across the full width of the plane, allowing them to cut into the corners of the joint. Shoulder planes are invaluable for trimming joints. With my 1-in. shoulder plane, I removed just the marks left by the bandsaw blade on the cheeks of the tenon and planed them smooth. That's all. When I put the joint together again, it fit perfectly. There is nothing terribly difficult about cutting a good mortise-and-tenon joint. It takes a little patience, some concentration, and an understanding of the mechanics of wood joinery. Remember: Whatever wood you remove from one piece, you must leave on the piece fitting to it.

When the mortise-and-tenon joint fits perfectly, you can turn your attention to the other end of the arm, which terminates in a lively scroll design. Drill out the inside circular segment with a sharp Forstner bit, then cut out the rest of the design on the bandsaw. Any jogs or blips can be cleaned up with a fine rasp and 60-grit sandpaper.

Two small brackets will support the mailbox shelf across its width. Make the brackets first, then cut two dadoes in the top of the arm to hold them. Cut the dadoes slightly narrower than the thickness of the brackets, then plane the sides of the brackets until a snug fit is achieved.

THE BRACE

The curved brace (see the drawing below) supports the arm from underneath. Start by gluing up two 20-in. pieces of 4x4 to obtain a workpiece wide enough to accommodate the curve of the brace. Don't use splines to join the pieces because you might cut through them later and mar the surface of the brace. The curve

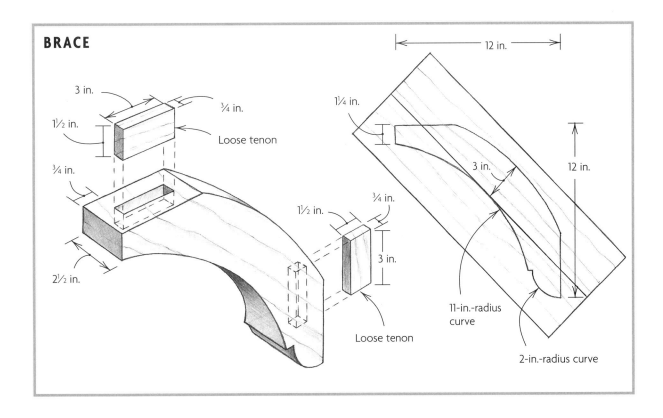

BRACE

3 in.

¾ in.

Loose tenon

1½ in.

¾ in.

2½ in.

1½ in.

3 in.

Loose tenon

12 in.

1¼ in.

3 in.

12 in.

¾ in.

11-in.-radius curve

2-in.-radius curve

for this post is based on an 11-in. radius. It's broad enough to provide good support, yet sharp enough to give it some real shape and interest. The brace extends 12 in. in each direction, along the post and the underside of the arm.

Before cutting the curve of the brace, resaw the material to 2½ in. in thickness. This gives the brace a more tailored and elegant look. After bandsawing the curve, clean up both the inside and outside curves with a spokeshave.

On a construction such as this, integral tenons would complicate the execution of the joinery and sacrifice some strength. Loose

tenons are a simpler solution because the ends of the brace can be cut to fit against the post and the arm without protruding tenons getting in the way.

On the brace, rout out the mortises on a router table, using a ¾-in.-dia. bit to a depth of ¾ in. On the post and arm, drill the mortises first with a ¾-in.-dia. Forstner bit to the same depth. Be sure to drill and rout the length of the mortise on the short side, then later open them up and square the corners with a chisel to accept the tenons.

The loose tenons are made of red oak, cut slightly oversized on the table saw and then planed to fit the mortises. Why red oak? Red

oak is a harder wood than redwood, and therefore it is less likely to deform during assembly and affect the fit of the joint.

THE SHELF

Don't put your mailbox on a single solid piece of wood; that would be an invitation to disaster—the wood would eventually split. Instead, the mailbox will sit on a two-piece shelf with a ³/₁₆-in. space in the middle and breadboard ends (see the drawing below), which will make a strong platform that won't twist or split with exposure to the weather. The shelf halves are joined to the breadboard ends with a ¾-in.-deep tongue-and-

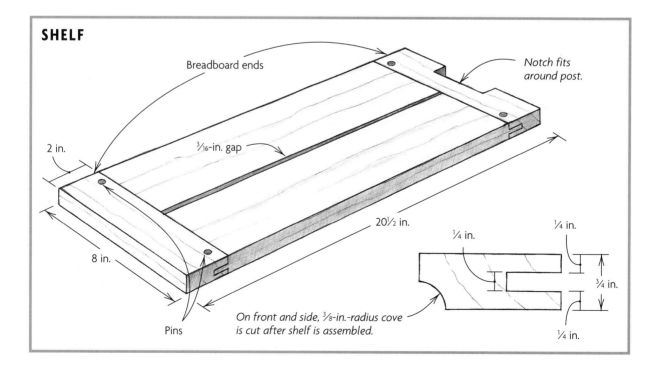

SHELF

Breadboard ends

Notch fits around post.

2 in.

³/₁₆-in. gap

20½ in.

8 in.

Pins

On front and side, ³/₈-in.-radius cove is cut after shelf is assembled.

¼ in.

¼ in.

¾ in.

¼ in.

groove joint, then pinned near the outside edges, allowing them to move freely along the breadboard edge. Since the wood is ¾ in. thick, the tongue is cut ¼ in. thick. The joint can be cut on the table saw.

The groove is cut first in the breadboard piece, then the tongue at the end of the shelf halves is cut to match. Begin by marking the center of the breadboard's edge. With the table-saw blade set for a ¾-in.-deep cut, make a ⅛-in.-wide kerf on each side of the marked line, yielding a groove that is ¼ in. wide. Next, mark the dimensions of the tongue off the groove (see the top photo at right). With the blade set to cut on the waste side of the tongue, pass the halves over the blade on end, cutting the tongue to thickness and length. Before you change the blade-height setting, check the tongue against the groove. If the fit is satisfactory, lay the panel flat on the table saw with one end against the blade and adjust the blade height to the underside of the tongue. Now, set the fence at ¾ in. from the left side of the blade and cut the shoulders of the tongues on the panels.

After the mailbox shelf is assembled, it must be notched to fit around the post (see the middle photo at right). To lighten the edge, cut a ⅜-in. cove on the underside of the shelf along

Check the end of the mailbox shelf panel against the grooved breadboard end before cutting the tongue.

The mailbox shelf is notched to fit around the post, then screwed to the arm.

A thin block raises the mailbox off the shelf enough for the door to swing open.

the sides and front edge. Then
screw the shelf to the arm.
To provide clearance for the
pivoting hinge of the mailbox
door (see the bottom photo
on p. 165), the mailbox must be
raised off the shelf about ⅜ in.
A solid block or a frame of
mitered strips can be used for
this purpose.

THE POST CAP

To make the cap (see the drawing at right), cut a 5¼-in.-sq.
block from a piece of 1-in.-thick
scrap, and rout a ½-in. cove on
the underside. On top of the cap,
plane a gentle slope so water will
run off easily. Make the flat on
the post cap large enough to
accommodate the dimensions of
the finial. Screw the post cap
into the post with two #8 screws,
placed so as not to interfere with
the attachment of the finial.

Beneath the post cap and around
the post, place a 2-in. skirt molding, beaded along the bottom
edge. This give the top of the
post some visual weight and bal-
ances it with the scrolled arm.
Miter the skirt molding strips at
the corners, then glue and nail
them in place.

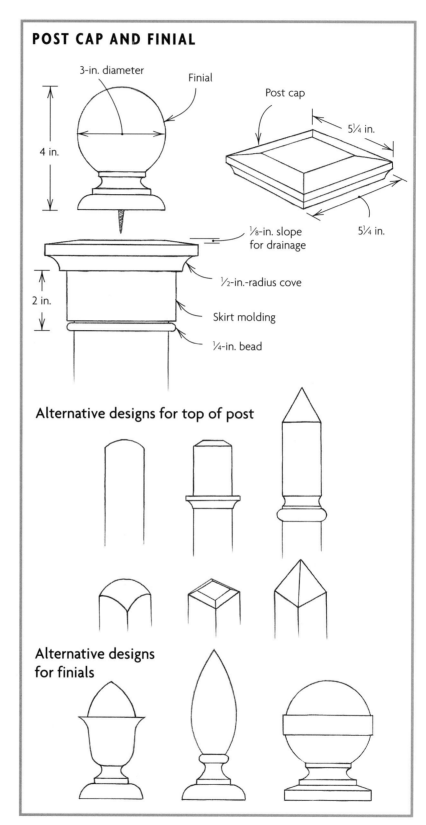

POST CAP AND FINIAL

3-in. diameter

Finial

Post cap

5¼ in.

4 in.

⅛-in. slope
for drainage

5¼ in.

½-in.-radius cove

2 in.

Skirt molding

¼-in. bead

Alternative designs for top of post

Alternative designs
for finials

ASSEMBLY AND INSTALLATION

The shelf and brackets are assembled with countersunk #8 stainless-steel screws. Only a few joints need glue—the tenon at the end of the arm, the loose tenons at either end of the brace—and for those, use a weatherproof glue such as Elmer's. After coating the joints, attach the brace to the underside of the arm, then join this assembly to the post and clamp everything together. After about 45 minutes drill a ⅜-in. dia. hole centered on the through tenon and pin the mortise and tenon with a dowel coated with glue. After the assembly dries, unclamp the post. For an authentic period touch, you could use the technique of drawboring (see the sidebar below), but it's not really necessary if you use glue.

Screw the finial into the post cap. Two finishing nails set through the base of the finial and into the post will prevent the finial from being removed easily.

Now the post is almost ready to set into the ground, but first you should coat the in-ground end with wood preservative. Three coats brushed on will protect the buried wood from rot and parasites. As for finish, if your post is cedar or redwood, a clear waterproof coating is fine. With exposure to the weather, it will age nicely and turn a beautiful silvery gray. If the post will be painted, prime with an exterior-grade primer, then sand, then apply two coats of exterior paint, followed by a third touch-up coat on the most exposed surfaces.

DRAWBORING

During the 17th and 18th centuries, mortise-and-tenon joints used in the construction of furniture and small interior architectural components such as doors, window sashes, and shutters were sometimes reinforced with pegs driven through the joint. Early adhesives were made from animal hides and bones and didn't have the strength or durability of modern glues and epoxies. Without pegs, the joint would eventually loosen as a result of glue failure, wood shrinkage, and seasonal movement.

In house framing, the giant joints used for the assembly of the timber frame were not glued, so pegs were driven into the joint to hold it together. Because the wood was unseasoned and exposed to extremes of hot and cold, simple pegging was not enough reinforcement. Drawboring refers to the practice of offsetting the holes drilled into the mortise and through the tenon for the purpose of drawing the parts tightly together when a pointed peg was driven into the joint (the point was trimmed off later). This practice was commonly employed by timber-frame housewrights to ensure the integrity of the joinery and the rigidity of the house frame.

INDEX

B

Baseboard:
  caulking, 54
  designing, 47-49
  installing, 51
  material for, 48-50
  one-piece vs. built-up, 47-49
  period, 47
Battens:
  edge treatments for, 39
  milling, 41
Brass, antiquing techniques for, 35

C

Casing:
  design for, 112-13
  installation of, 116-18
  material for, 112
  milling, 114-15
Catches, types of, 26
Caulk, as surface preparation, 31
Chair rail:
  decorative band for, 75-77
  designing, 74-75
  installing, 77-79
  one-piece, 80-81
  period, 73-74
  purpose of, 73
Chamfers, milling, 63-64
Coping, described, 52-53
Cove cuts, on table saw, 151-52
Cupboard, corner:
  assembly of, 154-57
  case construction of, 140, 155
  cornice for, 150, 151-52
  designing, 137-39
  face frame for, 140-41, 143
  fascia for, 155-56
  finish for, 157
  glazing for, 150
  installation of, 157
  lower doors for, 146-47, 157
  materials for, 139
  moldings for, 156
  in period houses, 137
  pilasters for, 143-44, 145, 155-56
  shelves for, 152-54
  upper door for, 147-50, 157
  waist molding for, 144-45, 146

D

Dado joint, discussed, 19
Design:
  modular: 85, 124
  period, sources for, 9-17
  See also specific projects.

Door, board-and-batten:
  designing, 37-38
  hanging, 43
  material for, 37-38
Door, frame-and-panel:
  assembling, 131-32
  design of, 123-24
  material for, 125-26
  milling, 126
  panels for, 124-25
  traditional vs. modern, 123
Dovetail joint, discussed, 19
Drawboring, discussed, 167

F

Flooring:
  edge treatments for, 61-64
  installing, 67-71
  layout of, 66
  low-luster finish for, 66
  milling, 61-63
  nailing, 66-67, 71
  oil stain for, 64-65
  ordering, 61
  polyurethane for, 65
  predrilling technique for, 66
  prefinished, 60
  saw for, 66, 67
  seasoning, 61
  shiplapped, 60
  See also Floors.
Floors:
  material for, 59-60
  in old houses, 59
  See also Flooring.
Fluting:
  on chair rail, 81
  by machine, 142
  molding planes for, 142-43
Framing:
  and chair-rail attachment, 78
  stud, 6-7
  timber, 6,7

G

Grooves:
  discussed, 19
  milling, 126

H

Half-lap joint, discussed, 18
Hardware:
  forged, 22-23
  See also specific items.

Hinges:
  antiquing technique for, 35
  for board-and-batten doors, 43
  butt, 26
  butterfly, 23, 25
  Euro-style (32mm), 26
  H, 24
  snipe, 23
  strap, 23
Historic American Buildings Survey
    (HABS), as design source,
    12-14
Houses:
  Colonial, 5-7
  historical sleuthing of, 14-17

J

Joinery. *See specific joints.*

K

Knobs, Colonial, 25

L

Locks, for doors and drawers, 26

M

Mailbox post:
  arm for, 161-63
  assembly of, 167
  brace for, 163-64
  cap for, 166
  design for, 159-60
  finial for, 166
  installing, 167
  materials for, 160
  post for, 160-61
  shelf for, 164-66
  stock preparation for, 160
Mantel:
  central panel for, 102
  cornice for, 102-103, 104
  design for, 97-98
  finish for, 109
  foundation for, 98-99
  frieze band for, 100-102
  installation of, 106
  pilasters for, 99-100
  secondary molding for, 103-104
  shelf for, 106-107, 109
Medium-density fiberboard (MDF), for
    interior woodwork, 22

Miters:
  with donkey ear, 108-109
  with miter jack, 108
  with shooting board, 108-109
  stepped, 120-21
  for wide moldings, 115-16
Molding:
  applied, 132
  built-up, 48-49
  for chair rail, milling, 77
  coped, 51-53
  cornice, 104-106
  cove, 104-105
  quirk-and-bead, 114-15
  shoe, 54
  traditional vs. modern, 111
  wide, miter for, 115-16
  See also Casing.
Mortise-and-tenon joint:
  for frame-and-panel door, 126-30
  discussed, 19
  fitting, 129-30
  haunched, 19, 127
  through wedged, 132

N

Nailer, pneumatic, for chair rail, 27,
  51-54, 79
Nails:
  air-powered, 27, 106
  cut, 25
  hand-forged, 25
  imitation cut, 25
  leather washers for, 42-43

P

Paint:
  antiquing, 31-32
  applying, 31
  crackle finish for, 33
  distressing, 34
  glazing, 33-34
  historical, 28
  latex, 29
  layered look for, 32
  milk, 28
  primer for, 31
  sponging, 32
  stippling, 32-33
  surface preparation for, 29-31
  texturing, 32
Panels:
  beading, 131
  for frame-and-panel door, 125-26
  milling, 130-31
  molding treatments for, 125
  reeded and notched, 133-35

Pine, for interior woodwork, 22
Planer, 12-in., 61-62
Planes:
  block, 50
  for fluting, 142-43
  jointer, 68-70
  miter, 108-109
  molding, 55-57
  moving fillister, 80, 134
  reeding, 134-35
  shoulder, 134, 162-63
  tongue-and-groove, 44-45
Plywood:
  edging for, 154
  for interior woodwork, 22

Q

Quirk and bead:
  with router, 41
  with scratch stock, 41

R

Rabbet joint, discussed, 19
Reeding, milling, 75
Restorations, historic, as design source,
  9-10

S

Sash joints:
  with cope-and-stick router bits,
    148-49
  traditional method for, 149
Saws:
  chop-, 50
  jeweler's, 50
Screws, for joining casework, 26-27
Shelters, dugout, 5,6
Shiplap joint:
  discussed, 18
  on floor boards, 50
Societies, historical, as design source,
  10-12
Staples, air-powered, 27

T

Tenons:
  on bandsaw, 128
  on table saw, 127
  wedged, 129
Thumb latches, 24-25

Tongue-and-groove joint:
  assembly of, 41
  discussed, 18
  with router, 39-40
  with tongue-and-groove planes, 39
  for window-seat lid, 92

W

Wallpaper, as wall finish, 27-28
Window seat:
  assembly of, 87-88
  base molding for, 94
  cleats for, 91
  designing, 83-84, 86
  fascia panel for, 88-89
  finger grip for, 95
  finish for, 94-95
  installation of, 90-91
  lid for, 91, 92-93
  materials for, 86
Windows, apron for, 118-19
Wood:
  plainsawn vs. quartersawn, 21
  See also Pine.
Woodwork:
  interior, 8
  See also specific projects.

Publisher: **JIM CHILDS**

Associate Publisher: **HELEN ALBERT**

Assistant Editor: **STROTHER PURDY**

Publishing Coordinator: **JOANNE RENNA**

Editor: **RUTH DOBSEVAGE**

Designer/Layout Artist: **LYNNE PHILLIPS**

Photographer, except where noted: **MARIO RODRIGUEZ**

Illustrator: **MARK SANT'ANGELO**

Typeface: **BERKELEY OLD STYLE ITC MEDIUM**

Paper: **68-LB. G-PRINT**

Printer: **QUEBECOR PRINTING, TENNESSEE BOOK OPERATIONS**

Additional photo credits

pgs. 4, 36, 46, 58, 72, 82, 110, 122,136, 158, **VIN GRECO**

p. 20, **SCOTT PHILLIPS**

p. 96, **BOYD HAGEN**